The Historiography of Contemporary Science and Technology

Studies in the History of Science, Technology and Medicine
edited by John Krige, CRHST, Paris, France.

Studies in the History of Science, Technology and Medicine aims to stimulate research in the field, concentrating on the twentieth century. It seeks to contribute to our understanding of science, technology and medicine as they are embedded in society, exploring the links between the subjects on the one hand and the cultural, economic, political and institutional contexts of their genesis and development on the other. Within this framework, and while not favouring any particular methodological approach, the series welcome studies which examine relations between science, technology, medicine and society in new ways e.g. the social construction of technologies, large technical systems.

Other titles in the series

Volume 1
Technological Change: Methods and Themes in the History of Technology
edited by Robert Fox

Volume 2
Technology Transfer out of Germany after 1945
edited by Matthias Judt & Burghard Ciesla

Volume 3
Entomology, Ecology and Agriculture: The Making of Scientific Careers in North America, 1885–1985
Paolo Palladino

Volume 4
The Historiography of Contemporary Science and Technology
edited by Thomas Söderqvist

Other Volumes in Preparation

Making Isotopes Matter: F.W. Aston and the Culture of Physics
Jeff Hughes

Molecularising Biology and Medicine: New Practices and Alliances, 1930s–1970s
Soraya de Chadarevian & Harmke Kamminga

The Historiography of Contemporary Science and Technology

Edited by
Thomas Söderqvist
Roskilde University, Denmark

harwood academic publishers
Australia • Canada • China • France • Germany • India • Japan
Luxembourg • Malaysia • The Netherlands • Russia • Singapore
Switzerland • Thailand • United Kingdom

Published in The Netherlands by Harwood Academic Publishers.

 Printed in India.

Amsteldijk 166
1st Floor
1079 LH Amsterdam
The Netherlands

British Library Cataloguing in Publication Data

The historiography of contemporary science and technology.
— (Studies in the history of science, technology and medicine ; v. 4)
1. Science — Historiography 2. Technology — Historiography
I. Söderqvist, Thomas
509

ISBN 3-7186-5906-9

Contents

Preface

Historians of science (including related areas of history of technology and history of medicine) are confronted with a paradox: more than 90 percent of all science in history has been produced during the last half century but so far only a small fraction of historical scholarship deals with this period. One reason for this glaring discrepancy may be that historians who approach the recent and contemporary technoscientific scene are confronted with new and unfamiliar methodological and theoretical problems. How shall we handle the huge amounts of published and unpublished sources? Is it possible to write a synthetic history of recent science? What level of scientific training is necessary to understand recent and contemporary science? Does the lack of historical distance prevent good scholarship? Can (and will) historians of recent science share the turf with other professional groups, such as active scientists, scholars of science and technology studies, and science journalists? How shall we deal with scientists' and technocrats' constant interference with our work? Whose history are we writing? Whose science?

These and many other questions were the focus of an International Workshop on the Historiography of Contemporary Science, Technology and Medicine, held on the 16–17 September 1994 at Göteborg University, Sweden. The workshop which was hosted by the Department of Theory of Science and Research was attended by around 40 participants from Europe and the United States. Most of the chapters in this volume are revised versions of papers presented at the workshop.

The thirteen chapters cover a great variety of topics, stretching from the role of participant observation in the recent history of immunology and interviews in recent biomedical science to the possibility of identifying a specific postmodern science. In the introductory chapter (a draft version of which was sent as a background paper to the participants before the conference), a number of methodological issues are discussed that confront historians of recent technoscience, e.g., the overload of published and unpublished source materials, the use of quantitative methods for writing synthetic historical narratives, how to handle the relation with the historical actors, and the problem of insufficient scientific training in writing about recent technoscience. Jeff Hughes questions the significance of some of these issues, particularly the idea of synthetic narratives of recent science, and adds another set of problems, including the fundamentally political character of historiographical choices. If there are problems unique to the history of recent and contemporary science, he concludes, they lie in the increasingly bitter politics of scientific legitimation.

Accordingly the contests surrounding recent and contemporary science include the sometimes vexed relations between historians and scientists. Susan Lindee maintains that what makes the history of recent and contemporary science a profoundly new kind of history is that it involves the historical actors in the process; a circumstance which raises, among other things, the problem of the historians' social responsibility towards the scientists. Soraya de Chadarevian also focuses on the tensions between historians and scientists and reviews the tradition of oral history of science against the background of her experiences of interacting with molecular biologists. Joseph Tatarewicz takes his experiences of writing the history of the Hubble Space Telescope Project as his point of departure for discussing the close ties between patron and historian: too much sponsor control leads to loss of credibility and the issue of control and credibility is thus continuously negotiated. Ilana Löwy, finally, describes her experiences as a participant observer in a biomedical laboratory and how the scientists' acceptance of her presence was based on the expectation that she would be a witness to the project's predicted success.

Memory — and its counterpart, forgetting — is another important theme in the historiography of recent and contemporary science. Drawing on his studies of the history of molecular biology (allosteric proteins), Jean-Paul Gaudillière analyzes how historical accounts told by scientists differ from those generated by other professional groups and the tension between the native production of memory and the professional writing of history. Skúli Sigurdsson draws on his work on a commissioned history of the electrification of Iceland and problematizes the habit of speaking in terms of scientific and technological progress, a major obstacle for writing the history of recent technoscience. Progress talk induces forgetfulness and enshrines the present at the cost of the past; historians of technoscience might let forgetting as well as remembering, be an integral part of their repertoire of self-authentication.

Another recurrent topic consists of the strategic problems involved in producing an overview of recent technoscience. Susan Cozzens reports on her work on the relation between co-citation maps of the 'research agenda' of recent neuroscience and the neuroscientists' own mental maps, and raises the question of whether the effort involved in creating synthetic overviews of major research areas is worthwhile. Frederic Holmes argues that the number of historians will never keep pace with the exponential growth of science and the differentiation of specialized research fields; the few historians of recent technoscience will therefore have to choose whether to spread their forces out evenly in the territory and thereby risk being isolated inquirers, or gather around a few emerging historical focal points, such as molecular biology, thereby further exaggerating the prominence of these fields of study.

Several chapters touch directly or indirectly on questions of moral responsibility. Paul Forman sees the 'moral economy' of late twentieth century science to be in many respects quite different from that of modern science since the 'scientific revolution'; he suggests that the instrumentalization and fragmentation of knowledge production in postmodernity puts the notion of responsibility squarely on the agenda of studies of technoscience and that historians must be prepared to confront thorny moral issues. Ronald Doel explores a new field in the study of recent and contemporary history, viz., the borderline between contemporary diplomatic history and the history of contemporary science. After a thorough literature review he discusses, among other things, how covert funding for international science policy affected peer review and decisions about the directions of major research programs during the Cold War.

Steve Fuller provides the volume with a summary of the main positions concerning the historian's role with respect to contemporary science; he then contrasts the epistemic roles played by historically based arguments in the natural sciences versus the humanities and the social sciences, and finally ends with an argument for a refurbished alternative grand narrative of the history of science.

A few words of acknowledgments. The participants of the two-day Göteborg workshop were a *sine qua non* for making this book possible. Their enthusiastic and knowledgeable comments provided an atmosphere of open intellectual exchange, and the many frank remarks greatly helped to improve the papers selected for this volume.

I would like to thank Svante Lindqvist for his support in organizing the workshop, Aant Elzinga and the Faculty of Humanities and Arts, Göteborg University, for providing us with a wonderful setting (Ågrenska Villan) for the meeting, and Alice Malmström and Anna Söderqvist for sorting out all possible practical problems during the two meeting days. The generous economic support of the Swedish Council for Research Planning is gratefully acknowledged.

I am also grateful to John Krige for being such a prudent series editor, to Lone Avlund Guldager for copy-editing, and Harwood Academic Publishers for their care in the production of the book. And finally, my most sincere thanks to the other twelve authors for their contributions, their cooperation and their patience.

Notes on Contributors

- **Soraya de Chadarevian** is a research fellow with the Wellcome Unit for the History of Medicine at the University of Cambridge. She has written on nineteenth and twentieth century biology and is currently working on a book about molecular biology in postwar Britain. She is editing *Molecularising Biology and Medicine* with Harmke Kamminga (forthcoming in this series).

- **Susan E. Cozzens** is professor at the Department of Science and Technology Studies at Rensselaer Polytechnic Institute. She has written about the history of contemporary neuroscience, including *Social Control and Multiple Discovery in Science: The Opiate Receptor Case* (1989).

- **Ronald E. Doel** is visiting assistant professor in the Department of History and the Geophysical Institute of the University of Alaska. He has published on the earth sciences and international science in the twentieth century, including *Solar System Astronomy in America: Communities, Patronage, and Interdisciplinary Research, 1920–1960* (1996).

- **Paul Forman** is a curator at the Smithsonian Institution in Washington, D.C. For the past thirty years he has written on the history of twentieth century physics and more recently also on historiography and postmodernity.

- **Steve Fuller** is professor in sociology and social policy at the University of Durham and editor of the journal *Social Epistemology*. He is the author of several books, including *Philosophy, Rhetoric, and the End of Knowledge* (1993). His latest book will be on the origins and impacts of Thomas S. Kuhn's book on the structure of scientific revolutions.

- **Jean-Paul Gaudillère** is a senior research officer at the Institute Nationale de la Santé et de la Recherche Médicale in Paris. He has worked on the history of recent molecular biology and his book *Le colibacille, le biologiste et le plan: histoire de la biologie moléculaire en France* is forthcoming.

- **Frederic L. Holmes** is Avalon Professor in history of medicine at Yale University. He has worked extensively on the history of eighteenth to twentieth century physiology and chemistry and is the author of several books, including *Hans Krebs* (2 vols., 1991–1993).

- **Jeff Hughes** is a lecturer at the Centre for the History of Science, Technology, and Medicine at Manchester University. He has worked on the history of nuclear physics and radioactivity; his book *Making Isotopes Matter: F.W. Aston and the Culture of Physics* is forthcoming in this series.

- **M. Susan Lindee** is associate professor at the Department of the History and Sociology of Science, University of Pennsylvania. Her books include *Suffering Made Real: American Science and the Survivors at Hiroshima* (1994) and *The DNA Mystique: The Gene as a Cultural Icon* (with Dorothy Nelkin, 1995).

- **Ilana Löwy** is a senior research officer at the Institute National de la Santé et de la Recherche Médicale in Paris. She has written about the history and sociology of twentieth century medicine and particularly the history of immunology, and has recently published *Between Bench and Bedside: Science, Healing and Interleukin-2 in a Cancer Ward* (1996).

- **Skúli Sigurdsson** is currently a research fellow at the Max Planck Institute for the History of Science in Berlin. He is, e.g., the author of *Herman Weyl, Mathematics and Physics, 1900–1927* (Ph.D. dissertation, Harvard University, 1991); his study of the electrification of Iceland is forthcoming.

- **Thomas Söderqvist** is senior lecturer at Roskilde University. He has written about various problems in the historiography of twentieth century science. His books include *The Ecologists: From Merry Naturalists to Saviours of the Nation* (1986) and a biography of the immunologist Niels K. Jerne (forthcoming).

- **Joseph N. Tatarewicz** has been a curator at the Smithsonian Institutions Air and Space Museum in Washington, D.C. and is now in private practice. He has worked and published on various aspects of space flight history for more than fifteen years, including *Space Technology and Planetary Astronomy* (1990).

CHAPTER 1

Who Will Sort out the Hundred or More Paul Ehrlichs? Remarks on the Historiography of Recent and Contemporary Technoscience

Thomas Söderqvist

A TEMPORAL IMBALANCE IN THE HISTORY OF SCIENCE

Only one or two decades ago most historians of science considered recent science — the scientific culture created, lived and remembered by contemporary scientists — an area of study best left to the historical actors themselves. The recent past was thought to be best fitted for scientists who tried to make sense of their immediate antecedents in autobiographies, textbooks, and review articles.[1] Professional historians of science, in contrast, were supposed to dig into the texts and contexts of the real and truly dead past, from antiquity to the early twentieth century.

This uneasiness with recent history probably has many roots, among them a wish to guard the professional status of the young discipline. If historians of science were to establish themselves as an independent community of scholars, immune to warnings against falling into the trap of adopting the actors' transcendent spirit and accusations of presentism, its practitioners had better stay away from recent and contemporary science. To study the history of molecular genetics, plate tectonics, NMR spectroscopy, or environmental controversies was considered somewhat suspect, verging on illustrated magazine journalism, dangerously close to partisanship, and inviting the bias which is said to accompany a lack of historical distance.[2]

These caveats notwithstanding, there are several good reasons for historians of science to direct their interest to the last decades. As Derek J. de Solla Price once pointed out, the number of scientists has grown exponentially during the last three hundred years.[3] The curve seems now to have reached a (temporary?) limit, but the overall exponential growth pattern nevertheless has an important implication, namely, that more than 90 percent of all scientists that ever existed in world history have been active since the end of World War II.[4] A similar pattern of chronological distribution can most likely be demonstrated for other indicators, including the number of publications, the number of scientific meetings, and the funds allocated to scientific research. (Lack of reliable statistics before the era of national research policy makes it difficult to underpin such estimates quantitatively.) It is difficult to avoid the impression that, by all measures, recent and contemporary technoscience is quantitatively at least on a par with all previous science. The fact that the bulk of scientific activity in world history has taken place in the last half century is in itself a simple but compelling motive for plunging into the recent history of science.

Has recent science's predominance in terms of workforce, publications and economic investment been paralleled by a corresponding wealth of cognitive development? In a sense probably not. It is hardly reasonable to assume that 90 percent of all major conceptual revolutions, epoch-making new techniques, institutional innovations and influential research schools in world history have seen the light since 1945, or that 90 percent of all Newtons, Priestleys, Darwins and Freuds are waiting out there in the recent dark to be identified, investigated and pantheonized by the historians of recent science; such quasi-quantitative accounts of cognitive developments are hardly meaningful. Indeed, some of the prodigious output of recent science may be attributed to repetition and redundance, for example, to the tendency to chop up research reports into several independent articles or to the fact that many research groups work simultaneously on fairly similar problems. It could also be argued that the really foundational contributions to scientific knowledge and the truly innovative institutional creations — the university, the academies, the laboratory, the research university, and so forth — were made in (much) earlier periods, so that post-1945 science is just 'more of the same'.

Yet the emergence of entirely new fields of inquiry, such as molecular biology, quantum chemistry and computer science, and of new powerful techniques, such as information technology, speaks against the claim that recent technoscience is just another half century accretion to the previous four centuries of cognitive and institutional renewal that has characterized modern science since the so called 'scientific revolution'. Furthermore, recent technoscience involves a significant proportion of women and engages scientists from nations and ethnic groups outside the traditional European-North-American core area. Large numbers of scientists are now employed in work-situations more similar to those of technicians than that of the classical autonomous scientist. Finally, new types of research organization, such as governmentally organized Big Science in the physical sciences and the increasing hybridization of university research and high-tech private enterprise in the life sciences, have appeared.[5] So, recent technoscience seems to be 'more of something new'.

In spite of the impressive growth and qualitative change of science over the last half century, however, historians of science still largely turn their attention to earlier historical periods. Thus a paradoxical temporal imbalance is created: on the one hand, the great majority of historians of science are still working on the small percentage of scientific activity that took place before 1945 (for example, British history of science graduate students still largely choose their dissertation topics from the seventeenth century through the first decades of the twentieth);[6] on the other hand, the bulk of science in world history has so far received the attention of only a small fraction of the historians of science. Such a conspicuous imbalance is not found in other historical fields: in political history, economic history, literary history, history of music, etc., the study of the recent past has given rise to specialties with their own journals.

Maybe the successful professionalization of the history of science over the recent decades has contributed to a more relaxed attitude to the alleged dangers of journalism and historical proximity — there are indications that the recent technoscientific system is about to get some of the attention its volume and permeation of global nature and culture deserves. The percentage of publications listed in *Isis Critical Bibliography* dealing with the twentieth century (i.e., since 1914) as a percentage of all chronologically arranged publications has risen from about ten percent in the early 1970s to about thirty percent in the late 1980s, an increase in scholarly interest that can hardly be accounted for by the expansion in time span only.[7] Unfortunately, no similar statistics exist for historical studies dealing with science since 1945; a fair guess is that around ten percent of the studies deal with the last fifty years.[8] Similar trends can probably be demonstrated in the history of recent technology and medicine.[9] The publication in 1992 of an issue of the journal *Osiris* dealing with science after 1940 epitomizes this new interest in recent history of science.[10] "Indeed", wrote Arnold Thackray, "science after '40 now remains as a sort of last frontier in the history of science — a frontier that is itself endless".[11] Maybe chronicling historical events or mapping 'white spots' of the past is not primarily what history should be about, but the prospect of exploring 'endless frontiers' nevertheless seems to be intrinsically attractive.

This orientation towards recent science is even more pronounced when we consider science studies: sociology of science, social studies of scientific knowledge (SSK), science policy studies, and so forth.[12] Studies of the social construction of scientific knowledge, of scientific controversies, of the rhetoric of science, of gender and science, etc., have regularly been set up as historical case-studies of recent science.[13] Although not necessarily addressing historical problems explicitly, many scholars in the field of social studies of science have nevertheless made contributions of interest for the history of recent science, thereby adding to the impression that the recent past is, at last, about to become a significant and legitimate part of history of science.

This growing interest in the history of recent and contemporary science puts a new question on the agenda, namely, how is this history to be written? One could of course argue that this is a superfluous question, since historians of recent science obviously have much in common with historians of other periods. Classical methodological issues, such as problems of periodization and the critical evaluation of written sources, are shared by all historians irrespective of the chosen time period. The advantages and drawbacks of rational reconstruction and sociological explanation are similar whether one works with medieval astronomy or recent molecular genetics. The issues of presentism vs. historicism and historical relativism, of particularism vs. contextualism, of how to relate structure with function, or how to integrate biography with institutional history are the same for the historian of pre-modern or modern science and for the historian of recent and contemporary science.[14]

But historians of recent science are also confronted with new and unfamiliar methodological obstacles which are not, or only rarely, shared by historians of earlier periods. I will discuss some of these obstacles in the following paragraphs.

THE PROBLEM OF DOCUMENTARY OVERLOAD

One of the obstacles confronting the historian of recent and contemporary science is the imbalance in the supply of primary sources. There is a noticeable scarcity of sources in archives and a corresponding overflow of sources outside the archives. Many documents from individual scientists, laboratories, research institutions, and scientific societies are indeed deposited in public or private archival institutions.[15] But these documents are not necessarily available: many archives have clauses on recent collections which only allow access after a certain (and often long) amount of time and, more important, they rarely contain the material searched for. The bulk of documents produced by recent technoscience are still in the custody of individual scientists. Practically all the documents from intermediary stages in the research process (letters, faxes, laboratory notes, instrument readings, manuscripts, etc.) are still to be found in desk drawers, on the shelves of the laboratories, and in the basements of departmental buildings.

The source imbalance problem is aggravated by the overflow of non-deposited documents. In addition to the mountains of papers piling up in the research institutions, historians of recent science have to consider the rapidly accumulating number of virtual documents in the form of electronic files on multimegabyte storage media. With respect to archival research, scholars of recent and contemporary science are thus confronted with a very different set of problems than scholars of earlier historical periods are used to coping with. Finding a new Robert Boyle manuscript makes the seventeenth century historian's day and even warrants an enthusiastic message in a newsletter;[16] likewise, if a couple of new clandestine Darwin notebooks were to be found they would surely provide for a life-long academic career. But loading another hundred thousand contemporary laboratory journal pages into the university archive or moving a couple of hundred megabytes to the historian of recent science's hard disk hardly makes his or her day; it either causes a severe headache or threatens to overload the local server.[17] The problem is not specific for technoscience — political historians and business historians confront similar problems when writing about contemporary state administrations and multinational firms — but it is severely aggravated the closer we come to the present and thus sets recent technoscience apart from earlier periods.

A related problem facing historians of recent and contemporary science is the seemingly ever-expanding production of scientific publications. The number of research articles abstracted each year in *Biological Abstracts* is now about 400,000; the number of entries in *Chemical Abstracts* well exceeds 700,000 per year. The number of articles in physics, mathematics, geoscience, electronics, medicine, etc., add up to around three million published documents a year.[18] The cumulation over the last fifty years is probably in the order of one hundred million research reports — a conservative estimate since patents and most industrial and military research reports are not included in the abstract journals.

QUANTITATIVE MAPPING AND SYNTHETIC EFFORTS

The overload of written sources is an acute problem for curators of history of science archives: which documents are worth keeping and which can be discarded? But it also creates problems for the historian of recent science: which documents and publications are significant and which constitute background noise? "There are dozens of Ehrlichs, Bordets, and Landsteiners around", complained the immunologist and Nobel laureate Niels K. Jerne when trying to write the history of recent developments in his discipline: "How then will the history of immunology of the last 25 years be written? Who will sort out the contributions of a hundred or more good scientists?", he asked before giving up the task.[19] Can we save all the papers of every scientist and every laboratory? Where is the limit to be drawn?

One solution to the overload problem would be to rely on the scientists'/informants' evaluation (in interviews or in review articles) of which documents and publications reflect important events and historical trends. But if historians of recent science do not want to be entirely dependent on the insiders' knowledge and experience, how then will we ever be able to orient ourselves in these forbiddingly vast amounts of published and unpublished documents?

The enormous publication output forces the historian to face the problem of quantitative methods and scientometric mapping of science. During the last decades most abstract journals and citations indexes have been provided through on-line databases (for example *Medline*) or, recently, on an increasing number of CD-ROMs. The historian of recent science is well advised to use these databases. *Chemical Abstracts* and *Biological Abstracts* began publication in 1907 and 1926 respectively and both have been available on-line for more than a quarter of a century. *Chemical Abstracts* can be searched on-line from 1967 to present and the 'CA old' file now covers the years 1957 through 1966. Similarly *Biosis* (*Biological Abstracts*) and *Science Citation Index* are available on-line from 1969 and from 1974 respectively. The Institute for Scientific Information in Philadelphia provides opportunities for searching the electronic *Science Citation Index* base retrospectively back to 1945. As a consequence most scientific texts published during the last fifty years are now available for citation analysis, co-citation analysis and other scientometric methods without much tedious manual work.

Historians of science in general have traditionally been very reluctant to use quantitative methods — and historians of recent science are no exception. For example, scientometric studies of citation and co-citation data have mainly been used for science policy purposes but rarely for historical studies of recent and contemporary science.[20] The reason for this hesitation on the side of historians may be very mundane, such as a lack of skill in using the new tools. Aesthetic reasons may also be an important inhibitive factor; old paper (at least before the introduction of acid paper in the late nineteenth century) is a more sensuous medium than the computer screen.

Another, and more principled, argument against the use of quantitative methods in the study of recent and contemporary science could be that these methods lend themselves primarily to the mapping of large-scale trends in scientific development, such as the development of research schools and disciplinary dynamics, and that the whole idea of such a synthetic overview could be seen as a remnant of an old-fashioned ideal, a corollary of a modernist ambition to map the world panoptically. There may be some truth in this: maybe the historiography of recent science is unavoidably stamped by the postmodern condition of fragmentation; maybe the dissolution of disciplinary boundaries of technoscience and the ever-increasing mutual infiltration of conceptual approaches and methodologies between research areas, specialties, and disciplines (which makes recent science look more like flexible post-Fordist production) negates the possibility of a quantitatively supported grand narrative of science since 1945, even grand narratives of particular scientific disciplines (or even research specialties).

Maybe the only possible, and desirable, history of recent and contemporary science consists in the narration of local histories, the critical analysis of theories, ideas and experiments rendered fashionable by stories in *New Scientist* and the editorial columns of *Nature* and *Science*, or the seemingly endless possibilities for permutation of new perspectives on the history of science: science and semiotics, science and gender, science and deconstruction, science and court culture, science and fractal mathematics, science and autopoiesis, and so forth. Maybe the history of recent science has to adopt the concept of the *Columbia Literary History of the United States* which "forgoes closure as well as consensus", whose 66 authors "acknowledge diversity, complexity, and contradiction by making them structural principles", and whose editors claim that it is no longer possible and desirable "to formulate an image of continuity".[21]

However, these and similar claims against an encyclopaedic ideal of historical writing — precluding not only quantitatively based syntheses but historical syntheses as such — are self-contradictory. If one acknowledges the ideals of diversity, complexity, and contradiction one has to accept a plurality of historical aims, including the wish to formulate unified and synthetic historical narratives. Historians who are not satisfied with contingent stories of spatially and temporally local episodes — who, like David Lindberg, have the ambition to make "synthetic efforts of every sort and at every level"[22] — cannot avoid drawing on large electronic databases when dealing with recent science.

Doing so raises new problems, however; problems common to quantitative studies in historical research in general, such as the proper use of statistical methods and cluster analysis, the textual and graphical representation of data, the problems of inference from large data sets to general patterns, and so forth. Furthermore, it raises the question of how the results of quantitative analysis can be accommodated with narrative history. Few historians of science would like to abandon the ambition to write stories which are readable, enjoyable and accessible to audience outside the circle of specialists. Narrative history goes hand in hand with 'the new contextual history', which stresses the complex interrelation between theories, institutional settings and larger social structures in ways that graphical representations or analytical discussions of the results of a quantitative analysis hardly can provide. Thus one of the major challenges that historians of recent and contemporary science have to deal with is how to write synthetic historical narratives without sacrificing either the comprehensiveness and 'objectivity' provided by quantitative indices or the 'soft' and richly textured contextual analysis.

HISTORICAL ACTORS AS INTERLOCUTORS AND MEMORY SOURCES

Scientometric analysis apart, the history of recent and contemporary science invariantly begins with the actors' own mundane oral history. History starts at the very moment when two scientists meet at the bench and discuss the outcome of yesterday's experiment. Shared stories begin to accumulate, rumours are communicated, myths become established. The historian of recent science arrives at a scene already heavily historicized. The arrival is not unproblematic, however. The fact that a large majority of the historical actors are still alive and willing to serve as interlocutors, memory sources and guides to the complex topography of the scientific landscape is yet another major challenge to our received notion of historiography of science.

The lack of written sources in the archives and the overflow of unlocalized and unregistered sources elsewhere makes oral history a *sine qua non* for those interested in understanding the development of recent science. Interviewing gradually became accepted in political and social history in the 1960s and soon made its way into the history of science; for example, one of the first histories of molecular biology was written largely on the basis of extensive interviews with the main, and then still active, actors.[23] A number of oral history projects have been established over the last decades. For example, the Center for History of Physics at The American Institute of Physics has long been active in the field; originating as a broadly conceived oral history project it now serves as a major repository of oral history interviews of American (physical) scientists. Many of these interviews are transcribed, indexed and accessible for historians. Several other similar projects and archives have recently been established. Consequently, an increasing number of historical studies of recent and contemporary science are now based, at least partly, on oral material.[24]

Oral history has its obvious strengths. Interviews with the historical actors can provide information — about personal motives, informal decisions, and hidden agendas — not available from written documents, they can support the interpretation of the significance of documents, and they can lead the historian to laboratory notebooks and private correspondence. But oral history is also permeated by problems which historians working on earlier periods do not have to encounter: the notorious unreliability of memory and the conflict between memory accounts and written documents,[25] the problem of interviewing across national and local scientific cultures,[26] the constructed and situated character of interview reports and the problem in transcribing phonetic accounts into writing.[27]

Some of these problems are particularly acute for historians who write biographies based wholly or partly on interviews with the subjects and their colleagues.[28] For example, so far all biographical studies of recent scientists have treated interview transcripts as just another kind of source of factual information without taking into account the fact that the life history is, at least partly, constructed in a specific social situation, namely, the interview. Biographical interviewing also raises the problem of the personal relation between the historian and the historical actor. Although the phenomena of emotional bonding and transference are most accentuated in biographical studies involving long-term personal co-operation,[29] their implications for oral history of recent science in general cannot be dismissed out of hand. How does the establishment of friendships between the historian and the scientists affect the outcome? How does the fact that many historians of science are recruited among relatively unsuccessful scientists color their attitude to science? These are little understood problems for the historiography of recent and contemporary science.

The practice of oral history also raises the problem of the validity of a purely cognitive approach to the history of science.[30] The fact that long-term biographical interviews allows the historian to see other sides of the personality than the official scientific mask, makes it difficult to sustain the traditional notion of scientists as primarily cognitive beings. The idea of the history of science as a detached study of the structure and dynamics of predominantly cognitive communities breaks down in biographical interview sessions; scientists instead emerge as passionate subjects with unique existential projects that transcend the abstractions of experimental systems and cognitively defined research communities.[31] The historian of recent and contemporary science is thus confronted with interlocutors who can support an existential understanding of recent science; an understanding that historians of earlier periods are largely denied because of the scarcity of appropriate written sources.

SCIENTISTS VS. HISTORIANS: CO-OPERATION OR 'SCIENCE WAR'?

Science in pre-modern and early modern eras was more seamlessly entwined with other social and cultural practices, ideas and institutions and therefore — by the very nature of its subject matter — its study claims more of cultural skills and the historian's ability to change his frame of interpretation than studies of present-day scientific knowledge. The need for cultural, hermeneutic and historical skills in writing about recent technoscience should, of course, not be underrated. But the voluminous production and diversity of the last fifty years makes the study of recent science much more technically demanding than writing about earlier periods;[32] so demanding that historians of recent science need systematic scientific training. This is true not only for highly mathematized fields, such as theoretical physics or cosmology, but also for the new life sciences with their technical and methodological complexity.[33] My own experience in writing about the history of recent immunology is that it is prohibitively difficult to evaluate the significance of experimental work in the field without either professional training or continuous and close co-operation with practicing immunologists.

Historians of pre-modern and modern science usually have some shoulders, however shaky, to stand on. The chances are high that other historians have already treated similar or related areas, if only as points of departure for designing one's research strategy and making preliminary interpretations. Historians of recent and contemporary science, however, rarely have any previous shoulders to stand on. Sometimes articles in specialized review journals and interviews with the actors-scientists can be used. But when entirely new vistas are to be explored — and remember that the history of recent and contemporary science consists mainly of entirely new frontiers — the field has to be mapped from scratch, and the historian has to stand face-to-face with the technical complexity of the science with no former historians at all to rely on.

Of course, isolated techniques, experimental systems or research programs can be mastered by non-scientists, for example by means of on-location training. Indeed a number of well-researched studies of recent and contemporary science have been written by non-scientists.[34] But historians with synthetic ambitions want to evaluate the importance of these techniques, experiments and research programs in their larger disciplinary and institutional context. For a richer understanding of, say, the preconditions of PCR-technology or the consequences of monoclonal antibody technology for recent science, a broad familiarity of the changing problem structure and institutional settings of the biomedical sciences of the last three decades is needed. One of the main reasons why historians have hesitated to deal with recent and contemporary science is probably that such familiarity is very difficult to obtain without systematic and professional scientific training. Obviously this is one of the major challenges to the future design of graduate programs for the training of new generations of historians of science.

But scientific training is not enough. With familiarity follows the dangers of dependence and partisanship. Historians professionally trained in science are always at risk of being carried away by the lure of the 'transcendent' attitude. The situation warrants great demands on the acquisition of historical, sociological, and ethnographic awareness. The historian of recent and contemporary science must, perhaps to a greater extent than historians of the more distant past, be scientist and historian in one person.

The inverted problem — the need for historical training and awareness — is not so often acknowledged in the history of recent science. Scientists usually have an ingrained respect for, even feelings of inferiority towards, historians working on earlier historical periods, particularly if it involves knowledge of the classical languages. In most scientists' view historians have a mystical ability to navigate among library card catalogues, localize unregistered documents in boundless archives, and produce texts without any visible writers' blocks. Scientists also acknowledges the fact that the reading of old texts demands specialized linguistic and even cultural interpretative skills, particularly if the science should be understood in the context of the age.

When it comes to recent and contemporary science, however, scientists often find it difficult to acknowledge the need for specific historical skills. Scientists sometimes find it hard to realize that the problems addressed by historians are of a different kind than those asked by the scientist: it is well-known that scientists, even when they are interested in times past, routinely ask about facticity and 'truth' at the expense of questions about the contextuality of science.[35] The scepticism is not entirely unfounded: the problems of interpretation in recent science are often more technical than cultural or linguistic; the work needed to disentangle experimental procedures and interpretation of data seems to overshadow the problems involved in revealing the metaphorical meaning of the technical vocabulary or disclosing the social and political context of laboratory work. It is therefore hard for scientists to accept that historical skills are nevertheless all-important also in the study of recent science, and that an historical awareness is not so easily obtained.

The problems mentioned here call for co-operation. But even if both historians and scientists remain respectfully within the boundaries of their professional roles, the co-operation is not without problems. All kinds of unpleasant problems are likely to crop up in the delicate balance between the aims of the historian and the aims of the historically interested scientist. This is not exclusively a problem of recent history, but it is only in dealing with recent science that the problem is acute, serious and seemingly unavoidable. Scientists are often unwilling to take the time and effort to try to understand the aims of historians, and often tend to believe that their main contribution to the co-operative effort is to 'set the historical record right' instead of seeing themselves in the role of witnesses and scientific experts. Conversely, historians may have so little understanding of recent science and may exhibit such insensitivity to scientists' way of thinking and understanding science that they cannot achieve the necessary rapport. Thus, the risk that historians 'go native' and lose their independence and critical function, is probably not so acute when we deal with recent science — it is not easy to forget on which side of the culture-gulf you are standing.

THE COLONIZATION OF THE HISTORY OF RECENT AND CONTEMPORARY SCIENCE

By focusing on events long past, historians of science have traditionally had very little competition from other professional groups; they have been used to having the turf for themselves, or peacefully sharing it with a few retired scientists, historians of ideas and philosophers of science. Turning to recent technoscience can be a frustrating experience because the historian now has to share the turf with new professional groups, including a growing number of active scientists, science journalists, sociologists of science, and scholars of science policy studies.

The dozens of recent Ehrlichs and their colleagues not only can, but often will, intervene in the writing of recent history. Long before the historians arrived, scientists have, of course, always defined the contemporary historical horizon by means of review articles, autobiographies, introductory chapters to textbooks, etc. All the way back to the birth of modern science its practitioners have shared stories about the origin and progress of their activities, honored their heroes and vilified their opponents. In this respect very little has changed. Today, like a hundred years ago, scientists write books in which they popularize scientific findings in the light of the recent past.[36]

During the last decade a growing number of professional science journalists have begun to explore the unfolding frontier of recent history of science to the effect that it is sometimes difficult to see the difference between high quality science journalism and well-written history of science;[37] journalists also tend to fare better on the tradebook market. The professional groups most active in the recent history of science, however, are probably sociologists of science and students of science and technology policy studies, who turn to recent and contemporary science to obtain material for more or less historically oriented case-studies.[38] All these groups feel equally called upon to write about recent science thereby setting professional conflicts and mutual accusations of 'naive realism' and 'higher superstition' right on the agenda.[39] Yet it is the collaboration and competition between these professional groups that makes recent and contemporary science such a vital and exciting topic for historical studies.

It is hardly necessary to point out that science and technology is a major — maybe *the* major — force in the shaping of the present world; it is also a truism to refer to the incompatibility between the two sets of Snow'ian cultures: those of the sciences and those of the humanities. Historians of science take on the task of contributing to a third culture: by means of a continuously unfolding frontier of inquiry historians of recent and contemporary science place themselves between those who change the conceptions of material world and those who — apologetically, ironically or critically — reflect upon these changing conceptions from a historical, sociological, or journalistic distance.

The historiography of recent science may thus become the focal point for a reflexive understanding of present technoscience. But reflexivity exercises are not enough. Most important the historiography of recent science may add some thoughtfulness and prudence to the technoscientific system, the prudence gained by years of accumulated experience. In the worlds of Aristotle:

> "Although the young may be experts in geometry and mathematics and similar branches of knowledge, we do not consider that a young man have Prudence. The reason is that Prudence [...] is derived from experience, which a young man does not possess; for experience is the fruit of years".[40]

In Aristotle's view prudence could be obtained between the age of forty and fifty. This is the individual age at which scientists used to begin to wonder about the history of their own discipline but also the collective age of professional history of science. Maybe it is about time for historians of science to co-operate with mellowed scientists to infuse some *phronesis* into the wildly galloping contemporary technoscientific system.

ACKNOWLEDGEMENTS

I am grateful to Ingemar Bohlin, Helge Kragh, Svante Lindqvist, Paolo Palladino, and Skúli Sigurdsson for comments on an earlier draft.

NOTES

1 I use the term science throughout for the natural sciences and their overlap with the technology and medicine. With respect to postwar events I alternate between the terms 'science' and 'technoscience'.
2 It should be remembered that suspicion against writing the history of (that time's) recent science has not always been the case. Nineteenth and early twentieth century historians of science (often scientists themselves and Whiggish to the core) saw no problem in dealing with their recent past. For example, Erik Nordenskiöld happily spent almost a quarter of his three volume survey of the history of biology on the last fifty years of biology, including contemporary frontiers such as cytology, neo-lamarckism, and experimental biology (Nordenskiöld 1928).
3 Price 1974.
4 In some areas maybe more: Bell Laboratories mathematician Andrew M. Odlyzko estimates that nearly half of the papers ever published in mathematics have appeared during the last decade (Stix 1994, 73).

5 For a discussion of postwar Big Science, see Galison and Hevly 1992.
6 See "A list of theses in history of science in British universities in progress or recently completed", edited by Sophie Forgan and regularly published by The British Society for the History of Science.
7 The figures are taken (with permission) from an unpublished manuscript by Paul Forman ("Independence, not transcendence, for the historians of science") given at Brandeis University in 1989. The figures were not published in his 1991 paper.
8 Among books from the last decade can be mentioned: Cozzens 1989, Gilbert 1991, Bud 1993, Aspray 1990, Hapke 1990, Close 1991, Bromberg 1991, Piore 1990, Habfast 1989, Le Grand 1988, Martin 1991, Heims 1991, Hermann *et al.* 1990, Mack 1990, Polkinghorne 1989, Wallace 1991, Brock 1990, Wills 1991, Nakayama 1991, Herman 1990, Kevles and Hood 1993, Glen 1994, and Levin and Brush 1995.
9 See, e.g., Grmek 1990 and Richards 1991.
10 *Osiris*, vol. 7, 1992 (edited by Arnold Thackray). The interest in recent science seems to be strongest in the US, while European historians of science have so far been somewhat more reluctant to go into the post-1945 period.
11 Thackray 1992, p. viii.
12 The field of science studies is coming of age, cf. Sheila Jasanoff *et al.* 1994.
13 See the journals *Science, Technology and Human Values, Social Studies of Science, Scientometrics*, etc.
14 Some standard problems in the historiography of science are summarized in Kragh 1989.
15 See, e.g., Warnow-Blewett 1992.
16 Hunter 1994.
17 And not only headache. The overload of documentation also affects the historian's way of understanding the conditions for intellectual production in a postmodern world. As Randall Collins says, "[t]o perceive the world as text is not too inaccurate a description, perhaps not of the world itself, but of the life position of intellectuals" (Collins 1992, 92). There are obvious practical and ethical consequences of this overproduction of cultural goods as Paul Forman has pointed out: overproduction produces "a crisis of impotence"; producers of intellectual goods have an obligation "to critically consider, and discriminatingly assimilate, the cultural 'goods on offer' " (Forman, this volume).
18 There is some overlap between these abstracts, for example, between *Index Medicus* and *Biological Abstracts* in biomedical fields.
19 Niels K. Jerne to Erna and Göran Möller, June 3, 1985 (copy in the Jerne papers, Royal Library, Copenhagen).
20 Cf. however, Cozzens (this volume). Significantly Cozzens considers herself a sociologist, not a historian.
21 Elliott *et al.* 1988, xiii and xxi; Olby *et al.* 1990 come close to this ideal.
22 Lindberg 1995, 66. Lindberg's evaluation of the need for synthetic efforts was addressed to the study of medieval science, but is certainly valid for all

periods in the history of science, and especially so for the last fifty years when the sheer size of global technoscience invites to fragmented efforts from the side of the historians.

23 Judson 1979.

24 Cf. de Chadarevian (this volume).

25 See, e.g., Kotre 1995. See also Lillian Hoddeson, "The conflicts of memories and documents: dilemmas and pragmatics of oral history", unpublished papers presented at the conference "Interviews in Writing the History of Recent Science", Stanford University, 28–30 April 1994.

26 Rosa Haritos, "Finding out: context and subtext in scientific controversy", unpublished paper presented at the conference "Interviews in Writing the History of Recent Science", Stanford University, 28–30 April 1994.

27 See, e.g., Mishler 1991.

28 E.g., Provine 1986, Holmes 1991 and 1993.

29 See, e.g., Moraitis 1985.

30 For a 'cognitive' approach to the history of science, see Nersessian 1995.

31 Thomas Söderqvist, "After the 200th Hour: The A/Effects of Long-Term Interviewing for Science Biography", unpublished paper presented at the conference "Interviews in Writing the History of Recent Science", Stanford University, 28–30 April 1994. For a discussion of 'existential biography', see Söderqvist 1996.

32 Caveat: Maybe the volume of research referred to above is largely a variation on rather few theoretical and methodological themes. In other words, maybe the seemingly unfathomable production is highly redundant with respect to new ideas, new theories, and new experimental methods? If this is the case, the demand for technical expertise lies more in being able to see through these variations and to reduce redundancy.

33 True, the technical complexity of recent science may be a relative phenomenon, a result of the fact that the contemporary horizon always seem to be more complicated than times long passed. Perhaps the individual sciences also differ with respect to technical difficulty: mathematics and physics being 'as difficult' as they have always been, whereas the recent life sciences stand out as technically more difficult to understand today, because they have become much less descriptive, much less phenomenal, much less common-sensical, and much more abstract and mathematized since 1945, partly a result of the rapid infusion of the physical and chemical sciences in biology and medicine.

34 See, e.g., Rabinow (1996) on PCR techniques in biomedical research.

35 A good example is provided by Larry Holmes' co-operation with Hans Krebs: see Holmes, 1991 and 1992.

36 See, e.g., Hawking 1988, Gilbert 1991.

37 A good example is Judson 1979, through which the author took the step from journalism to historical scholarship. See also, Gleick 1987 and 1992.

38 See, e.g., Balogh 1991, Smith 1990, and Wilkie 1991.

39 Cf. the discussion that has followed upon the publication of Gross and Levitt 1994.

40 Aristotle 1968, 349–350 [1142a].

BIBLIOGRAPHY

Aristotle, *The Nicomachean Ethics*, transl. H. Rackham (London: Heinemann, 1968).

Aspray, William, *John von Neumann and the Origins of Modern Computing* (Cambridge, MA: MIT Press, 1990).

Balogh, Brian, *Chain Reaction: Expert Debate and Public Participation in American Commercial Nuclear Power, 1945–1975* (New York: Cambridge University Press, 1991).

Brock, Thomas D., *The Emergence of Bacterial Genetics* (Cold Spring Harbor: Cold Spring Harbor Laboratory Press, 1990).

Bromberg, Joan Lisa, *The Laser in America, 1950–1970* (Cambridge, MA, MIT Press, 1991).

Bud, Robert, *The Use of Life: A History of Biotechnology* (New York: Cambridge University Press, 1993).

Close, Frank, *Too Hot to Handle: The Race for Cold Fusion* (Princeton, NJ: Princeton University Press, 1991).

Collins, Randall, "On the sociology of intellectual stagnation: the late twentieth century in perspective", pp. 73–96 in M. Featherstone (ed.), *Cultural Theory and Cultural Change* (London, SAGE, 1992).

Cozzens, Susan E., *Social Control and Multiple Discovery in Science: The Opiate Receptor Case* (Albany: State University New York Press, 1989).

Elliott, Emory (ed.), *Columbia Literary History of the United States* (New York: Columbia University Press, 1988).

Galison, Peter and Bruce Hevly (eds.), *Big Science: The Growth of Large-Scale Research* (Stanford: Stanford University Press, 1992).

Gilbert, Scott F. (ed.), *A Conceptual History of Modern Embryology* (New York: Plenum Press, 1991).

Gleick, James, *Chaos: Making a New Science* (New York: Viking, 1987).

Gleick, James, *Genius: The Life and Science of Richard Feynman* (New York: Pantheon 1992).

Glen, William (ed.), *Mass-Extinction Debates: How Science Works in a Crisis* (Stanford: Stanford University Press, 1994).

Grmek, Mirko D., *History of AIDS: Emergence and Origin of a Modern Pandemic* (Princeton, NJ: Princeton University Press, 1990).

Gross, Paul R. and Norman Levitt, *Higher Superstition: The Academic Left and Its Quarrel With Science* (Baltimore: John Hopkins University Press, 1994).

Habfast, Claus, *Grossforschung mit kleinen Teilchen: Das Deutsche Elektronen-Synchrotron DESY, 1956–1970* (Berlin: Springer Verlag, 1989).

Hapke, Thomas, *Die Zeitschrift für Physikalische Chemie: Hundert Jahre Wechselwirkung zwischen Fachwissenschaft, Kommunikationsmedium, und Gesellschaft* (Hertzberg: Bautz, 1990).

Hawking, Stephen, *A Brief History of Time* (New York: Bantam, 1988).

Heims, Steve Joshua, *The Cybernetics Group* (Cambridge: MA: MIT Press, 1991).
Herman, Robin, *Fusion: The Search for Endless Energy* (New York: Cambridge University Press, 1990).
Hermann Armin, John Krige, Ulrike Mersits, and Dominique Pestre, *History of CERN, Vol 2: Building and Running the Laboratory, 1954–1965* (Amsterdam: North-Holland, 1990).
Holmes, Frederic L., *Hans Krebs, Vol. 1: The Formation of a Scientific Life, 1990–1933* (New York: Oxford University Press, 1991).
Holmes, Frederic L., *Hans Krebs, Vol. 2: Architect of Intermediary Metabolism, 1933–1937* (New York: Oxford University Press, 1993).
Hunter, Michael, "A new Boyle find", *The British Society for the History of Science Newsletter*, nr. 45 (October 1994), 20–21.
Jasanoff, Sheila, Gerald E. Markle, James G. Peterson, and Trevor Pinch (eds.), *Handbook of Science and Technology Studies* (London: SAGE, 1994).
Judson, Horace F., *The Eighth Day of Creation* (New York: Simon and Schuster, 1979).
Kevles, Daniel J. and Leroy Hood (eds.), *The Code of Codes: Scientific and Social Issues in the Human Genome* (Cambridge, MA: Harvard University Press, 1993).
Kotre, John, *White Gloves: How We Create Ourselves Through Memory* (New York: Free Press, 1995).
Kragh, Helge, *An Introduction to the Historiography of Science* (Cambridge: Cambridge University Press, 1987).
Le Grand, H.E., *Drifting Continents and Shifting Theories: The Modern Revolution in Geology and Scientific Change* (New York: Cambridge University Press, 1988).
Levin, Aleksey E. and Stephen G. Brush (eds.), *The Origin of the Solar System: Soviet Research, 1925–1991* (New York: American Institute of Physics).
Lindberg, David C., "Medieval science and its religious context", *Osiris*, vol. 10 (1995), 61–79.
Mack, Pamela E., *Viewing the Earth: The Social Construction of the Landsat Satellite System* (Cambridge, MA: MIT Press, 1990).
Martin, Brian, *Scientific Knowledge in Controversy: The Social Dynamics of the Fluoridation Debate* (Albany: State University of New York Press, 1991).
Mishler, Elliot G., "Representing discourse: The rhetoric of transcription", *Journal of Narrative and Life History*, vol. 1 (1991), 255–280.
Moraitis, George, "The psychoanalyst's role in the biographer's quest for self-awareness", pp. 319–354 in S.H. Baron and C. Pletsch, *Introspection in Biography; The Biographer's Quest for Self-Awareness* (Hillsdale, NJ: Erlbaum, 1985).
Nakayama, Shigeru, *Science, Technology, and Society in Postwar Japan* (London: Kegan Paul, 1991).
Nersessian, Nancy, "Opening the black box: Cognitive science and the history of science", *Osiris*, vol. 10 (1995), 194–211.
Nordenskiold, Erik, *The History of Biology: A Survey* (New York: Knopf, 1928).
Olby, R.C., G.N. Cantor, J.R.R. Christie, and M.J.S. Hodge (eds.), *Companion to the History of Modern Science* (London: Routledge, 1990).
Piore, Emanuel, *Science and Academic Life in Transition* (New Brunswick, NJ: Transaction Publ. 1990).
Polkinghorne, John, *Rochester Roundabout: The Story of High Energy Physics* (New York: W.F. Freeman, 1989).
Price, Derek J. de Solla, *Science Since Babylon* (New York: Yale University Press, 1974).

Provine, W.B., *Sewall Wright and Evolutionary Biology* (Chicago: University of Chicago Press, 1986).

Rabinow, Paul, *Making PCR: A Story of Biotechnology* (Chicago: The University of Chicago Press, 1996).

Richards, Evelleen, *Vitamin C and Cancer: Medicine or Politics?* (New York: St Martin's Press, 1991).

Smith, Bruce L.R., *American Science Policy Since World War II* (Washington, D.C.: Brookings Inst., 1990).

Stix, Gary, "The speed of write", *Scientific American*, vol. 271 (1994), 72–77.

Söderqvist, Thomas, "Existential projects and existential choice in science: Science biography as an edifying genre", pp. 45–84 in: Richard Yeo and Michael Shortland (eds.), *Telling Lives: Studies of Scientific Biography* (Cambridge: Cambridge University Press, 1996).

Thackray, Arnold, "Preface", *Osiris*, vol. 7 (1992), vii–ix.

Wallace, Bruce, *Fifty Years of Genetic Load: An Odyssey* (Ithaca: Cornell University Press, 1991).

Warnow-Blewett, Joan, "Documenting recent science: progress and needs", *Osiris*, vol. 7 (1992), 267–298.

Wilkie, Tom, *British Science and Politics Since 1945* (Oxford: Basil Blackwell, 1991).

Wills, Christopher, *Exons, Introns, and Talking Genes: The Science Behind the Human Genome Project* (New York: Basic Books, 1991).

CHAPTER 2

Whigs, Prigs and Politics: Problems in the Historiography of Contemporary Science

Jeff Hughes

INTRODUCTION

There is always a certain scepticism as to the possibility of 'contemporary history'. Peter Hennessy, author of an award-winning historical study of post-war Britain, records, for example, a delightful conversation which took place soon after his appointment to the Professorship of Contemporary History at Queen Mary and Westfield College, London:

> "I was about to lunch with Sir Douglas Hague at the Athanaeum, when, in the bar, Eric Hobsbawm crossed the room to tell me how pleased he was to hear I was joining academic life. Professor Hobsbawm paused to tell his guest, Sir Isaiah Berlin, of the news of my appointment. [[] 'Good, good', said Sir Isaiah, 'contemporary history is the *most* interesting thing. In my time it used to be called *journalism, journalism*'".[1]

While the label itself may be problematic, however, neither the concept nor the practice of contemporary history is new.[2] Indeed, it has been noted that

> "if it is said — as historians sometimes say — that the idea of contemporary history is a newfangled notion introduced after 1918 to pander to the demands of a disillusioned public anxious to know what had gone wrong with the 'war to end all wars', it is not unfair to answer that what was newfangled was not a concept of history firmly anchored to the present but, on the contrary, the nineteenth century notion of history as something dedicated entirely to the past".[3]

Directly or indirectly, historians have always grappled with the present in their dealings with the past. For many, the present has been, and continues to be, the primary object of concern when they approach the past. Others have recognised the fundamental and inescapable role of the present in shaping representations of the past. History, in this sense, is always contemporary.[4]

Traditionally, the history of science, perhaps more than any other kind of history, has been as concerned with the present as with the past in just this way.[5] Is it not, after all, a fundamental tenet of the 'whig' approach which so long characterised the history of science that the past should be approached through its relationship to the present; that one should judge the efforts of past scientists in terms of their contributions to currently accepted knowledge; and that contemporary science, insofar as it is to be understood historically, is to be seen as the natural and inevitable culmination of past progress? Even when they bury themselves in the distant past, scientists and whig historians of science keep one eye firmly in the present.

If all history is, in some sense, contemporary, it is nevertheless true that the historical study of recent and contemporary events faces particular methodological and historiographical problems, for history, as A.J.P. Taylor put it, "gets thicker as it approaches recent times".[6] Many more traces of the recent past are available and the destruction and decay of documents and artefacts have hardly begun to set in, so that the contemporary historian tends to suffer from an embarrassment of riches as far as sources are concerned.

There is no doubt that contemporary and recent science, too, pose new and sometimes difficult methodological problems for the historian. Many of them are articulated in the chapters in this volume. The size and complexity of the archive for contemporary science, the theoretical and practical problems of oral history, the technical complexity and increasingly specialised character of recent science, the constraints imposed by secrecy, the demands of institutional patrons and many other factors combine to make the history of contemporary science one of the most interesting and challenging areas of the history of science today.

Yet it is possible to exaggerate both the novelty and the severity of these methodological difficulties and, in so doing, to neglect other, more general (and less often discussed) issues concerning the *politics* of history which might, ultimately, be of more significance both for the writing of contemporary scientific history and for the history of science as a discipline. In this chapter, I should like to comment briefly on some of the concerns expressed by Söderqvist and other contributors to this volume, with the aim of putting some of the methodological problems of the history of contemporary science into a somewhat broader perspective. I want to do so not in order to dismiss the practical difficulties and methodological problems which undoubtedly face contemporary historians, but in order to draw attention to a set of complementary issues which are equally pressing and which are perhaps, in the long run, of more importance for the historiography of science, past and present.

The roots of the issues I want to discuss lie in the way the history of science has developed as a discipline. For many years, the history of science was written largely by scientists themselves. For them, as Rachel Laudan has argued, the genre was an important part of "a programme of achieving a cultural niche for science based on [...] claims of epistemic preeminence".[7] Even those like George Sarton and Charles Singer who founded the academic discipline shared many of the assumptions and ambitions of the scientist-historians, and often saw the history of science as contributing to an overarching grand narrative of intellectual, moral and social progress. By the middle of the twentieth century, the history of science was an important buttress to the ideology of science.[8]

With the development of the discipline over the last two or three generations, however, a new attitude to science's past has emerged, and with it a range of different views of science's place in culture and society, past and present. Historians of science, approaching the past with a historical, rather than a scientific, sensibility have increasingly eschewed the diachronic, 'transcendental' approach of the scientist and the whig historian with its concern for the origins and development of great ideas and the allocation of proper credit to those individuals in the past who contributed to the development of the ideas which we today hold to be true. Drawing upon resources from anthropology, sociology and social theory, as well as from social, cultural and intellectual history more generally, historians of science have become increasingly interested in understanding not simply the products of past science as self-evident and independent entities, or the genesis and development of scientific thought, but rather the social and political processes by which knowledge was constituted in the past. The results of this exercise have been impressive. A wide range of historical studies now attest to the very different conditions under which past science was constituted, and we have finally begun to achieve a properly historical understanding of science.

To a significant extent, however, this shift from a broadly present-centred to a methodologically relativist approach to the history of science has had important implications for understandings of, and attitudes towards, the present, and indeed the future, of science itself. Unfortunately, it has also brought sociologists and historians of science increasingly into conflict with a number of scientists and historians who, for a variety of reasons, continue to see the history of science in its traditional mould as a linear, progressivist history of ideas. To take one example, a polemical 1987 article in *Nature* entitled "Whigs, prigs and historians of science" by Edward Harrison, Professor in the Department of Physics and Astronomy at the University of Massachussetts, warned the journal's (largely scientific) readership that "[i]n recent decades professional historians of science, trained in the social sciences, have established their own history of science societies, journals and meetings. Natural science and the study of its history", Harrison pointed out (apparently with some surprise), "have become divergent disciplines". He went on:

> "The whig interpretation of the history of science, practiced by most scientists according to historians, commits the crime of reconstructing past science in the context of today's science. The prig interpretation, practiced by many historians, adopts a superior attitude to historical work by scientists, and from fear of being unhistorical commits what it supposes to be the lesser crime of being unscientific".[9]

"By communicating the history of science only to historians", the article concluded, "and deigning not to communicate it to scientists, the historian lives neither in the past nor the present, [but] in a never-never land where ignorance is bliss, unable to evaluate (and if necessary to discount) the effect of modern science on modern styles of thought".[10]

Such remarks exemplify perfectly the problems to which I wish to draw attention in this essay. In recent years, a relatively small number of scientists, suspicious of historians' non-linear, non-whiggish and non-celebratory histories and of their increasing independence from science, see historians and those with whom they intellectually associate as somehow attacking the "moral stature", "epistemological authority", and therefore the legitimacy of science.[11] A debate has emerged in which recent work in history and sociology of science has been accused not only of undermining the legitimacy of science and its products and practitioners, but of attacking the very canons of objectivity and rationality which underpin the academy and indeed modern society.[12] This debate, I want to suggest, has important implications for the writing of history of science generally, and particularly for the writing of the history of contemporary science, because it is in the writing of contemporary history that the historian confronts the politics of the subject most directly through her attempts to secure, interpret and deploy evidence in the construction of historical accounts. History is always contested intellectual territory; for reasons to be explored in what follows, the history of contemporary science is *especially* contested.

ON METHODOLOGY, HISTORY AND CONTEMPORARY SCIENCE

Let me begin with some reflections on the methodological difficulties facing the historian of contemporary science, many of which, I suggest, are merely respecified versions of problems long familiar to historians, from which they differ only in degree, not in kind. Consider, for example, the 'problem' of the size and complexity of the archive looming before the contemporary historian. This has become so familiar a complaint as to have become almost a cliché. In many cases, indeed, the complaint is justified. In a recent foray into contemporary historiography, for example, Eric Hobsbawm notes the "unmanageable excess of primary sources" produced by "our endlessly bureaucratised, documented and enquiring times".[13] For the historian of science, too, the enormous expansion of the sciences in the post-war period, the accompanying social and intellectual prestige and authority which have made scientists more interesting to study in ever-greater detail, and the more recent introduction of new communication and data processing technologies — computers, fax, e-mail and so on — all mean that the range of sources available to the historian has vastly increased (not least, ironically, because many more scientists are preserving their papers to ensure their place in posterity).

But we should perhaps be careful not to confuse sometimes daunting complexity with fundamental difficulties. As Nathan Reingold rightly pointed out over a decade ago, "complaints about bulk often confuse what is at issue". "Bulk", he noted, "is a problem but not a new one".[14] Citing the 60,000-plus items in the Joseph Henry collection, and the enormous quantity of source material for the history of American Civil War in the U.S. National Archives and the Library of Congress, Reingold offers a salutary reminder that historians of every period have problems coping with the increasing amounts of potential data available to them. In the 1760s, for example, Joseph Priestley devoted more than half of his comprehensive survey of the science of electricity since earliest times to work done in the preceding two decades.[15] And a century ago, even Lord Acton ("whose mastery of written material", according to Seton-Watson, "was simply fabulous") could complain that there was "more fear of drowning than of drought" as far as sources were concerned.[16] In short, historians of all periods and at all times have faced the problems associated with the volume of source material.

Contemporary historians, it could be argued, face difficulties of a new order of magnitude. Yet there seems to be little evidence of this in practice. Political historians do not visibly crumple under the weight of archival documentation (even given the constraints of national security and the secrecy rules)![17] Nor do military historians wilt at the sight of the mounds of official documents and personal papers which routinely confront them at the Public Records Office and the Imperial War Museum! No: historians routinely deal with large archives, and it is sad but true that constraints other than bulk of material (for example, available research time, the constraints of official secrecy[18] or the limitations of one's photocopying budget!) usually intervene before sheer excess of material threatens to overwhelm the researcher. In practice, it must also be noted, archivists and historians of science are developing new methods for dealing both with this increased volume of material and with the new kinds of material finding their way into the historical record.[19]

For the (un?)lucky few for whom archival size really is a serious difficulty, the historian would typically *in practice* narrow the scope of her project precisely so as to make it manageable in the face of all the resources and constraints facing her. It is important to remember that, in any case, source material is invariably categorised according to the particular questions the historian wishes to address and the methodological and political imperatives she brings to bear in her research. Faced with a recalcitrant and unwieldy archive, set of documents or material artefacts, most historians would simply be more selective in their choice of material, and perhaps more explicit about the imperatives guiding that choice.

Complaints about size, then, can be greatly exaggerated. It may even be the case that historians — particularly contemporary historians — should worry more about the *destruction* of records than about unmanageable excesses. Harry Collins has rightly pointed out how the historical record can be shaped by the preservation of certain materials rather than others, as scientists (or, for that matter, historians) seek to ensure their place in posterity by preserving their papers. It is important to remember that the partial or wholesale destruction of records (often for equally self-interested reasons) can have an even larger impact on the historical record. In the 1970s, James Chadwick, heroic discoverer of the neutron in 1932 and subsequently leader of the British contingent in the Manhattan Project and a key player in post-war British science told Margaret Gowing, the official historian of the British atomic project, that "the whole idea of writing [such a] history was deplorable. It would, he said after a long pause, be much better to tear up all the papers".[20] Similarly, David Hounshell has recently pointed out some of the difficulties he and co-author John Kenly Smith, Jr. faced in writing their history of corporate research at DuPont, whose 1966 Records Retention Policy led to substantial record destruction and removed an enormous amount of valuable material from the archive, apparently the outcome of a somewhat arbitrary administrative decision within the company.[21]

These examples are salutary, for they illustrate that while there are particular methodological problems with the history of contemporary science, they are neither substantially different in kind from the problems that historians have always faced, nor do they seem to be insuperable: indeed, as the contributions to this volume indicate, we have gone a good way towards dealing with them effectively and efficiently in the everyday practice of our discipline.[22] Nor, moreover, are the most important methodological problems always to be found in the most obvious places.

That worries should be expressed about the size and complexity of the contemporary scientific archive may itself be indicative of certain underlying assumptions about the nature of the history of science. Given that historians routinely deal with huge quantities of material, for example, why might one ever have thought that the unmanageability of sources would be a problem for contemporary historians? Part of the answer, I suggest, lies in a particular view of the history of science as a totalising, synthetic account of the development of scientific disciplines, in which the role of the historian becomes that of adjudicator, sorting out the good science from the bad, the important contributions from the mundane and routine, and distributing credit appropriately. Given this view of the function of history, it is essential for the historian to master the archive and the discipline's literature in its totality for only by so doing can he produce the grand narrative constituting the History of a particular field.

Söderqvist affords a good example of this way of thinking. First, he seems to assume that the history of science is the history of scientific *ideas*, and that the proper task of the historian of science is to get to grips with those ideas, sort out the wheat from the chaff and craft a story linking them together and delineating the main lines of development from a broadly presentist perspective. Thus for Söderqvist the problem of technical difficulty in dealing with contemporary science is pronounced by the fact that historians of recent science not only want to understand isolated techniques, experimental systems or research programs" — they also want to evaluate the importance of these "in their larger disciplinary and institutional context".[23] Given this assumption, the sheer volume of material relating to recent and contemporary science can indeed become a serious difficulty, and only the individual with a complete mastery of the technical complexities of the science can possibly hope to contribute to the history of the field. While attention to the technical content of science is, and must remain a centrally important part of the history of science, however, this notion of history as somehow adjudicating over past science should not go unchallenged, for it seems to me to raise important questions about the role of the historian and of historical writing about the sciences.

A second, and I think even more serious assumption implicit in Söderqvist's historiography centres around the idea of completeness, the notion that it is *in principle possible* to write a complete and definitive history of a particular discipline, so that the relevant literature has to be grasped and understood in its entirety. Over a decade ago, Paul Forman denounced Jagdish Mehra and Helmut Rechenberg for attempting to produce a "complete account" of the emergence and development of quantum mechanics giving the "full story of all significant problems and their interplay".[24] What they had ended up with, complained Forman, was five hefty volumes of unstructured and often repetitive narrative, a mere catalogue of "one damn thing after another". "What", he ended up asking, "is history?" and "[w]hat is the task of the historian?"[25] *Pace* Söderqvist, this is where we too need to begin to reflect on what the history of science is and to whom it speaks, for the basis of Forman's criticism is that history is not, and cannot be, a complete record of past events. Because of the way history is constructed, moreover, it cannot be an objective and value-free record of events.

All historians recognise that the writing of history necessarily involves the selection and interpretation of traces from the past (or, indeed, the present). Crucially, however, when we write history, we "select a version of the past and a way of appropriating it that aligns [us] with some readings (readers) and against others".[26] In the case of the history of science, historians often compete with scientists in their interpretations of past science. Each group requires different things of its history, however, and the histories they produce embody different assumptions and different values. In this connection, Ashplant and Wilson take us to the heart of the matter when they observe that "all historiography rests upon acts of choice. No historiography can ever be a neutral enterprise: an enquiry into any given aspect of the past necessarily derives from some evaluation in the present. And, in the end, or rather in the beginning, all such choices are not simply historiographic, but political".[27] For the historian of (contemporary) science, as we shall now see, these choices are *especially* political.

THE HISTORY OF SCIENCE AS CONTESTED TERRITORY

History is always positioned. From Thucydices' apologetics to current debates about, say, Winston Churchill as a war leader, the relationship between Shostakovich, his music and the Soviet state or the history of AIDS, accounts of the distant and recent past — histories — are written *by* particular people *with* particular interests *in* particular settings *for* particular purposes. As Keith Jenkins puts it, "history is never for itself; it is always *for someone*".[28]

Who, then, is the history of science 'for'? In many respects, the history of science is different from many other branches of history in that both its practitioners and its principal audience have until relatively recently consisted mainly of scientists themselves. More than that, however, the history of science has traditionally played an important role in the legitimation of science. For reasons we (and, one suspects, they) understand perfectly well, for example, scientists tend to write largely celebratory, teleological and/or anecdotal accounts of the past. History, for the scientist, serves as a sustaining resource, legitimating (or delegitimating, according to which side of orthodoxy the particular scientist-historian happens to be) the *status quo*, establishing a temporal frame for current projects, providing exemplars, and providing a guide to future action.[29] As Forman puts it, "For scientists history is not the field upon which they wrestle for truth, but principally their field of celebration and self-congratulation" — potentially a problem, because "[h]istory written in celebration of heroic ages and agents can scarcely avoid becoming propaganda pandering to the scientists' *amour propre*".[30]

We must recognise, of course, that scientists write many different kinds of history, ranging from the brief and simplistic historical introductions to textbooks and literature surveys through reminiscences and biographical memoirs to full-blown (and often methodologically sophisticated) historical papers and monographs. It would be a mistake to denounce all these historical productions altogether as whiggish, celebratory and unreflexive: many scientists are only too well-aware of the contingencies implicit in historical writing.[31] Just as historians disagree about interpretations, for example, scientists themselves can disagree about the contours and interpretation of their *own*, apparently shared history.[32] Nevertheless, scientists and historians tend to find different things interesting about the past, to want to use their history for different purposes, and to select their sources and write their accounts accordingly.

For his part, Edward Harrison (author of "Whigs, prigs and historians of science") readily admits that scientists routinely use history to legitimate their current beliefs. "Traditionally", he notes "scientists have considerable interest in the history of science. When physicists seek to justify the modern quest for unified interactions, they naturally turn to the age of Maxwell where the quest began with the unification of electricity and magnetism". Indeed, history is a crucial resource for scientists not just in terms of the politics of *ex post facto* legitimation but in the very constitution of scientific knowledge at the laboratory bench itself:

> "In scientific research, where ideas form and dissolve in a state of flux and at any moment present countless potential futures, scientists retain their bearings by contrasting past and present ideas. Awareness of temporal depth in science forms an inte-

gral part of scientific research [...] Scientists grope their way, seeking to divine where they are going from where they are coming: they reach into the future as well as the past, and their diachronically extrapolated conceptual networks are not whiggish sins but the essence of science in action".[33]

History also serves a valuable pedagogical role in science, of course. In a celebrated 1974 article, "Should the History of Science be Rated X?" Stephen G. Brush pointed out the disjuncture between historians' and scientists' histories of science. Linear, teleological histories, he acknowledged, perform an important instructional role in science, and the science teacher "who wants to indoctrinate his students in the traditional role of the scientist as a neutral fact-finder should not use historical materials of the kind now being prepared by historians of science".[34]

That history serves as an important legitimating resource for scientists is not, in itself, a major problem: history finds meaning, after all, in its use.[35] The key problem for the historian, it could be argued, lies in the danger of conflict between the scientist's version of history of science and the historian's, especially where such conflict reflects (or is perceived as reflecting) antithetical social or political values concerning the status of science. Clearly, this danger is especially marked for the contemporary historian for, as Elizabeth Fee and Daniel Fox put it:

"In dealing with the distant past, historians have only the dead and each other with whom to contest their interpretations; in dealing with the recent past and the present, they must also confront the living — who have memories of their experience, and who may also have powerful and perhaps partisan explanations of the same events. The political and ideological struggles over interpretations of the present are usually waged with a special intensity rarely dispayed in arguments over the more distant past".[36]

Nathan Reingold formulates the problem even more succinctly. "While few living scientists will man the barricades over interpretations of the phlogiston theory", he notes, "many individuals can be cut close to the bone by contemporary history".[37] It is here, in the land of the living, not in the archive, that the problems of contemporary scientific history really lie.

In addressing the question of the relationship between historians and scientists, there are two principal issues of concern. First, there is the issue of contested interpretation and the difficulty of grounding historical analysis in the face of what might be a well-entrenched actors' history (and, indeed, in the face of potentially litigious actors). On the whole it must be said that scientists welcome historians' interest and are perfectly happy to give historians a free hand in their selection and deployment of source materials. But there are those scientists who wish to retain such control over their history that they will not tolerate anything that departs from the 'official' (heroic/celebratory/whiggish?) line. This can have serious consequences for the work of the historian, not least because historians often rely on the good offices of scientists for access to source materials. There is always the possibility that the historian's access to individuals and sources might be impeded if it is felt that the historian is writing 'inappropriate' history. More often that not, this is a question of delicate negotiation and compromise between historian and scientist. But as Joan Bromberg notes, for example, "historians of modern physics have pusillanimity as an occupational hazard. We depend upon the good graces of the scientists we treat for our raw material — the interviews, access to archival and personal collections, even, occasionally, our funding. The temptation to treat scientists with kid gloves is enormous".[38]

In the vast majority of cases, such difficulties never arise, or are carefully negotiated as part of the formal and informal relationship between historian and scientist or institution. Most scientists, to repeat, have enormous respect for the professional integrity and intellectual autonomy of historians of science. They recognise that historians and scientists have very different purposes in writing history, and are usually only too happy to cooperate in the matter of access to personal papers and other source materials. There are some, however, who do not take such a view. For them, the professional integrity and intellectual autonomy of the historian of science are threats rather than virtues. Indeed, some of the scientists for whom history serves such a valuable legitimating function consider themselves to have proprietary rights over the history of science to such a degree that they would even deny that any history written by non-scientists properly belongs to the history of science at all. For some scientists, moreover, history is so valuable a resource that to write history which doesn't legitimate science in some way is actually seen as positively *delegitimating* — in other words, as 'undermining' science in some sense — which can generate a profound hostility towards professional historians of science and their writings.[39]

For some, such as Harrison, this hostility towards historians of science derives from irritation at historians' self-righteous denunciations of scientists' habitual whiggery. Here the scientists may have a point, for anti-whiggism is a rhetorical card that can be overplayed. In other cases, hostility comes more from a frustration at historians' occasional reluctance to engage fully with the technical content of science.[40] In the last few years, however, a number of writings by historians of science have come under public and increasingly vitriolic attack from several quarters for the kind of history which they propound. In many cases, it must be said, these attacks are directed not at historians *per se* but at sociologists and other social scientists. Nevertheless, in so far as many historians of science today find useful resources in the sociology and anthropology of science, the attacks must be seen as being directed at historians of science also.

In his book *The Unnatural Nature of Science*, for instance, Lewis Wolpert, embryologist and a leading spokesman for the Public Understanding of Science movement in Britain, asserts that it is "unfashionable among historians of science to take what Herbert Butterfield called a Whig view of history — to interpret the past in terms of progress, as opposed to seeing it as series of events that have no particular direction. But it is in precisely this respect that science [...] is special: for the history of science *is* one of progress, of increased understanding".[41] Wolpert goes on to make a slashing attack on sociologists, philosophers and historians of science for their collective failure to uphold the standards and values of scientific rationality, and to suggest that what they should really be studying are not the processes by which knowledge is created and constituted but, for example, "what institutional structures most favour scientific advance, what determines choice of science as a career, how science should best be funded, how interdisciplinary research can be encouraged".[42] Ruling more fundamental questions out of court, Wolpert's remarks illustrate something of the political debate now beginning to shroud the history of science and cognate fields. While Wolpert's views by no means necessarily represent those of the majority of scientists, they should nevertheless be taken seriously as representing the public position of an influential and vocal group of self-appointed 'defenders of science' against the 'encroachments' of sociologists, historians and other relativist bogeymen from the social sciences.[43]

The reasons for this frantic boundary work invite further exploration. At a conference on 'The Social Standing of Science' at Durham University in December 1994, the scientists present dismissed the sociology of scientific knowledge and, by association, sociologically-informed history of science — contemporary and long past — on the grounds that such writings had contributed nothing substantive either to science or to our understanding of it as a social phenomenon![44] Why, though, should history of science as an academic discipline any longer have anything to contribute to science? Some light can perhaps be shed on the question by Jenkins' shrewd analysis of the role of groups of different status in the production of history:

> "The fact that history *per se* is an ideological construct means that it is constantly being re-worked and re-ordered by all those who are variously affected by power relationships; because the dominated as well as the dominant also have their versions of the past to legitimate their practices, versions which have to be excluded as improper from any place on the agenda of the dominant discourse. In that sense re-orderings of the messages to be delivered [...] just have to be constructed continuously because the needs of the dominant/subordinate are constantly being re-worked in the real world as they seek to mobilise people(s) in support of their interests. History is forged in such conflict and clearly these conflicting needs for history impinge upon the debates (struggle for ownership) as to what history is".[45]

It could be argued that a struggle is going on to define what the history of science is and what it should be. This, in places, is turning out to be a struggle to define what *science* is, and what *it* should be. Such considerations draw our attention inexorably towards the wider politics of the relationship between scientists and their would-be historians. For this relationship has changed beyond measure in the last few years. William Beik, for example, notes that "[in] a world deeply influenced by the collapse of the Communist regimes of eastern Europe and the reverberations of the conservative policies of the era of Ronald Reagan and Margaret Thatcher, it is perhaps understandable that historians are redirecting their attentions towards ruling élites, free-market mechanisms, and the independent influence of ideas and individuals".[46] For scientists, too, conservative administrations in the 1980s and the "end of the Cold War" have occasioned a (not always welcome) reappraisal of values and priorities, leading to what many have seen as a crisis of legitimation in the sciences, best exemplified perhaps by the cancellation of the Superconducting Supercollider.[47] A time when resources are being spread increasingly thin is no time for critical voices to be pointing out the contingencies and historical specificities constituing present science.

This politically charged atmosphere creates both opportunities and dangers for historians of contemporary science. There is the ever-present risk, for example, of the historian being caught in the crossfire of contemporary scientific controversy, and perhaps even of being enrolled as an ally in the debate. Similarly, despite Roy Porter's caution that the task of the historian of science is no longer "to play historiographer royal to science, but to undertake detached analysis of how science really operates, and to examine its place within the wider spheres of thought, culture and society",[48] the historian can find herself implicitly or explicitly cast in the role of 'official historian'. In the complex economy of the history of science today, in which contract work and commissions play an important role, the relationship between historian and patron must be carefully specified so as to avoid difficulties of this kind.

David Cantor has nicely analysed the politics of writing commissioned histories of contemporary medical institutions, for instance, and concludes that neither historians nor patrons emerge as well as they might from such exercises. 'Official' historians, constrained by the real and imagined needs and wishes of their patron, often pull their punches and produce pseudo-celebratory histories which toe the patron's line. More to the point, however, he sees such historians as confusing the distinction between science and history, producing seamless texts in which "[a]ctor and author, historian and scientist, witness and participant, flow into each other, so that any sense of causal agency is avoided".[49] This "blurring of the boundaries of history and science", he notes, "hides the political voice of scientists, while at the same time enabling them to assert their own interpretations of past events under the guise of history, and behind the mask of a professional historian".[50]

Echoing Forman's clarion-call for the moral, as well as intellectual and professional independence of the historian of science,[51] Cantor notes that "it is the historian's responsibility to maintain boundaries between his or her story and those of historical actors". "Obviously", he continues, "such a distinction becomes increasingly problematic as historians approach the present", but this is an issue to be explored and engaged with, rather than one to be shied away from. The history of contemporary science is precisely about negotiating this difficult terrain. It is clear that historians of contemporary science have to be able to reach some kind of accommodation with the scientists and institutions they write about, for, as we have seen, they often control rights of access to personal and professional papers and other resources essential to the historian's trade. But if, as Cantor suggests, this means following a particular "line" in the history we write, then clearly we have to think very hard about where our priorities lie.

CONCLUSION BUT NOT CLOSURE: THE FUTURE PAST(S) OF SCIENCE PRESENT

Cantor's remarks bring us back to the more general conflict emerging between those who seek to preserve the history of science as a grand narrative of progress and utility and those who seek to write a more *critical*, differently *engaged* and, so it could be argued, more *relevant* kind of history of science. Some have seen the 'new' history of science as potentially contributing to wider debates about the place of science in culture (for example to reshaping the public understanding of science movement, perhaps even to "democratising" science and improving its accountability, whatever that might mean).[52] At a time when communitarianism has been under attack and radical political and social individualism held up as the way forward, those who reject the politics of individualism have sought to use their writing to make the point that societies and social groups are important — not just for the production, validation and circulation of knowledge, but in other spheres of human activity too. In a parallel debate in social history, political issues are being explicitly identified by the participants as contributing to current trends in historical writing. The point of the post-1960s social history project we, recall, was to give voice to those whose contributions to history had been effaced or ignored by historians of "high" culture and politics.[53] In this sense, as Robert M. Young points out, the history of science "belongs in the mainstream of social and cultural debate".[54]

Within the contested terrain that presently constitutes the history of science, then, ought we, perhaps, to be arguing for what Allan Megill calls a "critical pluralism" in respect of the way we analyse science historically?[55] Easier said than done, perhaps, for the real problem, as Cantor rightly points out, is that academic historians (though not, it seems, scientists) try to shy away from this involvement, instead of embracing it. It is easy — too easy — for the historian to continue defending the professional and cognitive authority of science. But is it the historian's job to legitimate science and scientists, any more than it is necessarily the political historian's job to legitimate particular political systems, parties or doctrines? An uncomfortable question, perhaps, but one which the historian must face.

If there are problems unique to the history of contemporary science, they lie in the contests which increasingly surround the history of science and in the increasingly bitter politics of scientific legitimation. These are debates which historians of science cannot escape. In view of the fact that their work brings contemporary historians into day-to-day contact with scientists who are themselves situated with respect to particular debates and controversies within and around science, the question is vital.

ACKNOWLEDGMENTS

This paper was originally written in response to Thomas Söderqvist's position paper for the International Workshop on the Historiography of Contemporary Science, Technology, and Medicine held in Göteborg, Sweden, September 1994. I am indebted to him for this provocation, and for many agreeable historiographical discussions which have almost invariably ended in our agreeing to disagree. I also thank Deborah Brunton, Jon Agar, Svante Lindqvist, John Pickstone, and participants at the Göteborg Workshop for incisive comments on earlier versions of this paper.

NOTES

1 Hennessy 1994, 16.
2 See, e.g., Seton-Watson 1929.
3 Barraclough 1967, 15.
4 See Barraclough 1967, 15; Appleby, Hunt, and Jacob 1994, 52–90; Novick 1988.
5 In what follows, I shall largely be concerned with the history and historiography of science, and that mainly in the Anglo-American tradition. Parts of my argument will also apply to some extent to the history of medicine. In virtue of its very different intellectual roots, institutional forms and socio-political uses, the history of technology would perhaps make an interesting comparative case.
6 Taylor 1965, 602. On the methodological and historiographical problems associated with the study of recent and contemporary history in general, see Barraclough 1967; Seldon (ed.) 1988a, 1988b.
7 See Laudan 1993, 1; Schaffer, 1988.
8 Laudan 1993; Pyenson 1993; Reingold 1986. Also see Appleby, Hunt, and Jacob 1994.
9 Harrison 1987, 213–14.
10 Ibid.
11 Brush 1995, 224.
12 Key texts here are Wolpert 1992; Holton 1993; Appleby, Hunt, and Jacob 1994; Gross and Levitt 1994. For a cogent analysis of the debate from a science studies perspective, see Fuller 1994.
13 Hobsbawm 1993, 17.
14 Reingold 1990, 344.
15 Schaffer 1988, 73.
16 Quoted in Seton-Watson 1929, 7.
17 See Seldon (ed.) 1988; Hobsbawm 1993.
18 Here, Hobsbawm (1993, 17) notes that "the historian of his own times is not worse off than the historians of the sixteenth century but better off [be-

cause] at least we know what might, and in most cases sooner or later will, become available, whereas the gaps in the past record are almost certainly permanent".

19 Warnow-Blewett 1992; Lewenstein1995. The recent trend towards what has been called 'Big History of Science' exemplifies perfectly how historians of science are finding ways of coping with increasing complexity. For the term, see Gingras 1993; for examples, see Hermann *et al.* 1987; Hoddeson *et al.* (eds.) 1992; Hoddeson *et al.* 1993.

20 Gowing 1993, 80.

21 At the same time, however, Hounshell stresses the ambivalent character of document destruction: at one level, the quality of the book could have been improved had there been high-quality manuscript material on which to draw; at another, had no records been destroyed, "we would either still be working on this book or would have become so overwhelmed by the process that we would have given up". See Hounshell 1990, 405. Cf. Collins 1981.

22 De Hart 1993, 582, notes that while she shares many historians' preference for "a well-stocked manuscript room with its ease of access and aura of quiet detachment", the experience of working on contemporary topics posed "far fewer problems of perspective, sources, and evidence than is generally assumed".

23 Söderqvist, this volume.

24 Forman 1983, 824, quoting Mehra and Rechenberg.

25 Forman 1983, 825.

26 Jenkins 1991, 70.

27 Ashplant and Wilson 1988, 274.

28 Jenkins 1991, 17.

29 On the legitimating role of scientific histories, see Pickering 1989; Graham *et al.* (eds.) 1983. The recent cold fusion episode provided several examples of delegitimating history. Compare, for example, the interpretations of the controversy set out by Peat 1989; Close 1990; Mallove 1991; Huizenga 1993; Taubes 1993. See also Lewenstein 1992.

30 Forman 1983, 826.

31 Taylor 1981.

32 Witness, e.g., the debates among nuclear physicists in Wiener 1972.

33 Harrison 1987, 214.

34 Brush 1974, 1170.

35 Jenkins 1991, 26.

36 Fee and Fox 1989, 307.

37 Reingold 1981, 281.

38 Quoted in Brush 1995, 225.

39 Paul Forman, for example, has noted scientists' habitual contempt for, and refusal to engage with, many professional historians' writings on science. See Forman 1991, 82 (note 37). For a compelling analysis of two scientists' recent ripostes to historical and sociological "attacks" on the integrity of

science (together with some suggestions for what we might do about it), see Fuller 1994.

40 Brush 1995, esp. 226–228; Taylor 1981.
41 For this, and more in the same vein, see Wolpert 1992, 100.
42 Ibid., 122.
43 Cf. Midgley 1992.
44 Atkins 1995; Christie 1995; Fuller 1995; Johnson 1995. For a more general elaboration of this 'critique', see Wolpert 1992.
45 Jenkins 1991, 17–18.
46 Beik 1993, 207.
47 See Schmitt 1994.
48 Porter 1988, 71.
49 Cantor 1992. Cf. Hounshell 1990.
50 Cantor 1992, 134.
51 Forman 1991.
52 See, for example, Turner 1990; Fuller 1995.
53 For an excellent discussion of the history and politics of social history, see Wilson 1991, 9–58. See also the various contributions to the debate on "The Dilemma of Popular History," in *Past and Present* 1991, nr. 4.
54 Young 1988, 77.
55 Megill 1989, 653.

BIBLIOGRAPHY

Ankersmit, F.R., "Historiography and postmodernism", *History and Theory*, vol. 28 (1989), 137–153.

Appleby, Joyce, Lynn Hunt, and Margaret Jacob, *Telling the Truth About History* (New York and London: W.W. Norton, 1994).

Ashplant, T.G. and Adrian Wilson, "Present-centred history and the problem of historical knowledge", *Historical Journal*, vol. 31 (1988), 253–274.

Atkins, Peter, "Science as truth", *History of the Human Sciences*, vol. 8 (1995), 97–102.

Barraclough, Geoffrey, *An Introduction to Contemporary History* (Harmondsworth: Penguin, 1967).

Beik, W., "Debate: the dilemma of popular history", *Past and Present*, vol. 141 (1993), 207–215.

Berridge, Virginia and Philip Strong, "AIDS and the relevance of history", *Social History of Medicine*, vol. 4 (1991), 129–138.

Berridge, Virginia and Philip Strong (eds.), *AIDS and Contemporary History* (Cambridge: Cambridge University Press, 1993).

Brush, Stephen G., "Should the history of science be rated X?" *Science*, vol. 183 (1974), 1164–1172.

Brush, Stephen G., "Scientists as historians", *Osiris*, vol. 10 (1995), 215–231.

Cantor, David, "Contracting cancer? The politics of commissioned histories", *Social History of Medicine*, vol. 5 (1992), 131–142.

Capshew, James H. and Karen A. Rader, "Big Science: price to the present", *Osiris*, vol. 7 (1992), 3–25.
Christie, John, "The social standing of science: some contemporary history", *History of the Human Sciences*, vol. 8 (1995), 103–108.
Close, Frank, *Too Hot to Handle: The Race for Cold Fusion* (London: W.H. Allen, 1990).
Collins, Harry, "Understanding science", *Fundamenta Scientiae*, vol. 2 (1981), 367–380.
De Hart, Jane, "Oral sources and contemporary history: dispelling old assumptions", *Journal of American History*, vol. 80 (1993), 582–595.
Fee, Elizabeth and Daniel M. Fox, "The contemporary historiography of AIDS", *Journal of Social History*, vol. 23 (1989), 303–314.
Forman, Paul, "A venture in writing history", *Science*, vol. 220 (1983), 824–827.
Forman, Paul, "Independence, not transcendence, for the historian of science", *Isis*, vol. 82 (1991), 71–86.
Fuller, Steve, "Can science studies be spoken in a civil tongue?" *Social Studies of Science*, vol. 24 (1994), 143–168.
Fuller, Steve, "On the motives for the new sociology of science", *History of the Human Sciences*, vol. 8 (1995), 117–124.
Gardiner, Juliet (ed.), *What is History Today?* (Basingstoke: Macmillan, 1988).
Gingras, Yves, "Redefinitions in physics", *Science*, vol. 260 (1993), 1165–1166.
Gowing, Margaret, "James Chadwick and the atomic bomb", *Notes and Records of the Royal Society of London*, vol. 47 (1993), 79–92.
Graham, Loren *et al.* (eds.), *Functions and Uses of Disciplinary Histories* (Dordrecht: Reidel, 1983).
Gross, Paul and Norman Levitt, *Higher Superstition: The Academic Left and Its Quarrels with Science* (Baltimore: Johns Hopkins University Press, 1994)
Hamerow, Theodore S., "The bureaucratization of history", *American Historical Review*, vol. 94 (1989), 654–660.
Harrison, E., "Whigs, prigs and historians of science", *Nature*, vol. 329 (1987), 213–214.
Hennessy, Peter, "The pleasures and pains of contemporary history", *History Today*, vol. 44 (3) (1994), 16–17.
Hermann, Armin, J. Krige, U. Mersits and D. Pestre, *History of CERN* (Amsterdam: North-Holland, 1987, 2 volumes).
Hobsbawm, Eric, *The Present as History: Writing the History of One's Own Times* (London: University of London, 1993).
Hoddeson, Lillian, E. Braun, J. Teichmann, and S. Weart (eds.), *Out of the Crystal Maze. Chapters from the History of Solid State Physics* (Oxford: Oxford University Press, 1992).
Hoddeson, Lillian, P.W. Henriksen, R.A. Meade, and C. Westfall, *Critical Assembly. A Technical History of Los Alamos During the Oppenheimer Years, 1943–1945* (Cambridge: Cambridge University Press, 1993).
Hounshell, David and John Kenly Smith Jr., *Science and Corporate Strategy: DuPont R&D, 1902–1980* (Cambridge: Cambridge University Press, 1988).
Hounshell, David, "Interpreting the history of industrial research and development: the case of E.I. du Pont de Nemours & Co.", *Proceedings of the American Philosophical Society*, vol. 134 (1990), 387–407.
Holton, Gerald, *Science and Anti-Science* (Cambridge, MA: Harvard University Press, 1993).

Huizenga, John R., *Cold Fusion. The Scientific Fiasco of the Century* (New York and Oxford: Oxford University Press, 1993).

Jenkins, Keith, *Re-Thinking History* (London: Routledge, 1991).

Johnson, Julie, "Science friction", *New Statesman and Society* (13 January 1995) 29–30.

Judson, Horace F. and I.R. Mackay, "History in the Bay of Naples", *Immunology Today*, vol. 13 (1992), 459–461.

Laudan, Rachel, "Histories of the sciences and their uses: a review to 1913", *History of Science*, vol. 31 (1993), 1–34.

Lewenstein, Bruce V., "Cold fusion and hot history", *Osiris*, vol. 7 (1992), 135–163.

Lewenstein, Bruce V., "Do public electronic boards help create scientific knowledge? The cold fusion case", *Science, Technology and Human Values*, vol. 20 (1995), 123–149.

Mallove, Eugene F., *Fire from Ice. Searching for the Truth Behind the Cold Fusion Furor* (New York and Chichester: John Wiley, 1991).

Megill, Allan, "Recounting the past: 'description', explanation, and narrative in historiography", *American Historical Review*, vol. 94 (1989), 627–653.

Midgley, Mary, *Science as Salvation. A Modern Myth and its Meaning* (London: Routledge, 1992).

Novick, Peter, *That Noble Dream: The "Objectivity Question" and the American Historical Profession* (Cambridge: Cambridge University Press, 1988).

Peat, F. David, *Cold Fusion. The Making of a Scientific Controversy* (Chicago: Contemporary Books, 1989).

Pickering, Andrew, "Editing and epistemology: three accounts of the discovery of the weak neutral current", pp. 217–232 in L. Hargens *et al.* (eds.), *Knowledge and Society: Studies in the Sociology of Science Past and Present*, vol. 8 (Greenwich, HCT and London: JAI Press, 1989).

Porter, Roy, "What is the history of science?" in *What is History Today?* ed. J. Gardiner (London: Macmillan, 1988), 69–71.

Pyenson, Lewis, "Prerogatives of European intellect: historians of science and the promotion of Western civilization", *History of Science*, vol. 31 (1993), 289–315.

Reingold, Nathan, "Science, scientists and historians", *History of Science*, vol. 19 (1981), 274–283.

Reingold, Nathan, "History of science today, 1. Uniformity as hidden diversity: history of science in the United States, 1920–1940", *British Journal for the History of Science*, vol. 19 (1986), 243–262.

Reingold, Nathan, "Tales from the archives", *Proceedings of the American Philosophical Society*, vol. 134 (1990), 340–348.

Restivo, Sal, "Modern science as a social problem", *Social Problems*, vol. 35 (1988), 206–225.

Rosenberg, Charles, "Woods or trees? Ideas and actors in the history of science", *Isis*, vol. 79 (1988), 565–570.

Schaffer, Simon, "What is the history of science?" pp. 73–75 in J. Gardiner (ed.) "*What is History Today?*" (London: Macmillan, 1988).

Schmitt, Roland W., "Public support of science: searching for harmony", *Physics Today*, vol. 47 (1)(1994), 29–33.

Seldon, Anthony (ed.), *Contemporary History: Practice and Method* (Oxford: Blackwell, 1988a).

Seldon, Anthony (ed.), *The Longman Guide to Sources in Contemporary British History* (London: Longman, 2 volumes, 1988b).

Seton-Watson, R.W., "A plea for the study of contemporary history", *History*, vol. 14 (1929), 1–18.

Söderqvist, Thomas, "How to write the recent history of immunology: is the time really ripe for a narrative synthesis?" *Immunology Today*, vol. 14 (1993), 565–568.

Söderqvist, Thomas and Arthur Silverstein, "Participation in scientific meetings: a new prosopographical approach to the disciplinary history of science — the case of immunology, 1951–72", *Social Studies of Science*, vol. 24 (1994), 513–548.

Taubes, Gary, *Bad Science: The Short Life and Weird Times of Cold Fusion* (New York: Random House, 1993).

Taylor, Alan J.P., *English History, 1914–1945* (Oxford: Oxford University Press, 1965).

Taylor, Lauriston S., "Technical accuracy in historical writing", *Health Physics*, vol. 40 (1981), 595–599.

Turner, Joseph, "Democratizing science: a humble proposal", *Science, Technology, and Human Values*, vol. 15 (1990), 336–359.

Wallace, Mike, "The battle of the Enola Gay", *Radical Historians' Newsletter* nr. 72 (May 1995), 1–32.

Warnow-Blewett, Joan, "Documenting recent science: progress and needs", *Osiris*, vol. 7 (1992), 267–298.

Weiner, Charles (ed.), *Exploring the History of Nuclear Physics* (New York: American Institute of Physics, 1972).

Wilson, Adrian and T.G. Ashplant, "Whig history and present-centred history", *Historical Journal*, vol. 30 (1988), 1–16.

Wilson, Adrian, "A critical portrait of social history", pp. 9–58 in Adrian Wilson (ed.), *Rethinking Social History* (Manchester: Manchester University Press, 1991).

Wolpert, Lewis, *The Unnatural Nature of Science* (London: Faber and Faber, 1992).

Young, Robert M., "What is the history of science?", pp. 75–77 in J. Gardiner (ed.), *What is History Today?* (Macmillan, 1988).

CHAPTER 3

The Conversation: History and History as it Happens

M. Susan Lindee

INTRODUCTION: A CONVERSATION BETWEEN PAST AND PRESENT

History is generally about people who cannot terminate our employment contracts, write bitter letters to *Science*, or produce book-length critiques of our field.[1] Despite these advantages of working on the dead, historians of science are increasingly drawn to contemporary topics involving living persons. Historical study of post-war science and technology is almost fashionable and some historians now find themselves engaged in an extra-disciplinary activity, history as it happens.[2] This refers in this instance to the studies of academic historians who apply historical judgments and methods to contemporary events. It is not journalism, not anthropology, not sociology, and not exactly archival history — leaving open the question of what it is — but it has a growing corps of practitioners and an ascendant respectability that is in some cases earned.[3]

The British historian and philosopher R.G. Collingwood proclaimed in 1946 that all history is contemporary history for all history is a rethinking of the past in terms that bring it into the present. "The past is not known", said Collingwood, "but only the present".[4] Lucien Febvre noted in 1949 that history "gathers, classifies and groups together facts about the past in terms of its current needs. It interrogates the dead in terms of the living".[5] In the United States this relativist conception of the historian's craft has dominated academic circles since the 1950s. Its implications were fully articulated in Richard Hofstadter's *The Progressive Historians*, a text still routinely assigned to fledgling historians in American graduate schools.[6] Despite the rise in the 1980s of a 'new historicism' — a new emphasis on the importance of interpreting history in actors' terms — American historians continue, by and large, to expect history to shed light on contemporary concerns.

The notion that history is a conversation between past and present has particular cogency for historians of science in the 1990s, for contemporary science intrudes ferociously into everyday life, defining possibilities and shaping social choices. Historians of science are of necessity either covertly or openly attuned to contemporary science for contemporary science is the subject of history of science, just as contemporary gender relations are the subject of feminist history and contemporary political trends are the subject of political history. If the historian of science adopts the pose that contemporary concerns are irrelevant to the study of the past, that historian is reacting defensively to their suffocating, overwhelming relevance.

But what of those who embrace this relevance, and either forsake earlier centuries to study the late twentieth or explicitly use their historical studies to comment on contemporary science? Let me suggest some possible motivations for the historian of science to attempt this. The most intellectually respectable motivation is perhaps the conviction that it would be blind or stubborn to turn one's back on contemporary science. To ignore modern science, in this view, would be a willful refusal to recognize its importance to the history of science. Some historians are relatively overt about the contemporary implications of their work. Steven Shapin, in his *Social History of Truth*, for example, delineates the role of trust relationships and class in seventeenth-century England, and then closes his book with a critical commentary on their continuing importance (and relative invisibility) in contemporary science. Trust relationships, he claims, persist as central to truth in scientific core sets, small groups of specialists who know each other and can decide based on their assessments of individuals what data can be accepted as true.[7]

Similarly historians who work on the role of gender in science have been attentive to the implications of their work for the contemporary participation of women in science. One of the great ironies of the story of gender and science is that modern science — with its ideological origins in ideals of masculinity — has played a role in changing the social status and participation of women in the public world of work. Modern science began as an endeavor that freed men from the bodily constraints of subjectivity and transience and at the same time entrapped women more completely in their bodies. The missed opportunity of the rise of modern science, as Londa Schiebinger has put it, was that women did not benefit from the Cartesian dichotomy of body and mind.[8] Now modern science seems important to the systematic involvement of women in the world outside the home. Science and technology are not responsible for the changing status of women but they have participated (sometimes ironically) in those changes.[9] I mean to suggest that for many practicing historians of science their work is important precisely because it might help us understand this wonderful, terrifying thing we have made, modern science.

A more mundane justification for a historian's attention to contemporary science is that it promises employment. Academic jobs in the history of science are relatively rare and competition is intense. But there are jobs in federal labs, in private industry, in Washington agencies. In addition, there are funding sources, such as the National Center for Human Genome Research's Ethical, Legal and Social Issues program, committed to supporting studies of contemporary science, often with an interest in policy implications.[10] There may also be more public interest in studies of contemporary science. Many academics like the idea of reaching a broader audience even if they dislike the compromises involved. Historians of science who turn their attention to issues that appear in the day's headlines may be drawn partly to the opportunity to have a social impact. My own work with Dorothy Nelkin on images of the gene in popular culture was partly motivated by such concerns. The project began after we shared our concerns about the increasing use of genetic images in popular magazines, newspapers, advertising and fiction. In our study, we looked at earlier periods — at the popular eugenics literature of the 1920s and 1930s, and at mass circulation constructions of environmental determinism in the 1950s — but presented this historical material as a way to understand or explain contemporary concerns.[11] The project had an evangelistic quality: We hoped that by demonstrating the peculiar powers that DNA seemed to possess in popular culture we could help to put those powers in perspective. We intended to reach a broad audience and our emphasis on practical contemporary concerns seemed to make that more likely.

There is a third possible motivation for an attention to recent or contemporary science by historians: It is documented in a virtually infinite data source. While it is true that many fine records are available for nineteenth and even eighteenth century science and true that even scholars of the sixteenth or seventeenth centuries may find it difficult to examine all the possible evidence at their disposal, it is also true that there are more and more varied primary records of post-World War II science. Contemporary science is documented in the United States in the records of federal agencies overseeing patents, drug trials, health research, science funding and the nuclear energy program; in professionalized corporate archives; in personal papers and oral histories and autobiographies by leading American scientists. This mass of data is either a problem (as Söderqvist suggests in this volume) or a delightful mess. The historian's puzzle-solving instincts can be gratified by the excess of documents and living witnesses, by the conflicting interpretations, by the muddiness of the narrative.

My study of a post-war biomedical research program had just these qualities. There were many narratives, many sources, and living participants who had an active stake in my text. In the process of constructing this history — as I tried to understand detail, and then to reach beyond it — I had many options for triangulation. I could call participants to discuss particular questions; I could use published primary sources written by participants who had already constructed narratives of their own; and I could draw on records and correspondence preserved in private and public archives.[12] At the same time, of course, I learned more than I wished to know about the social complexity of writing histories in which historical actors are alive and, occasionally, hostile. It is this social dimension that interests me here.

Having just suggested why writing the history of contemporary and recent science is legitimate I want now to explore why it is so difficult to do. I suggest that despite all the attractions of doing contemporary history there is a single point against it that threatens to outweigh all those in its favor: Contemporary history involves living human beings. Doing contemporary history places one in social relationships that carry with them all the familiar benefits and risks of friendship, intellectual growth, hostility, revenge, and so on. The historian of contemporary science can wound and be wounded in ways that the historian of eighteenth-century science cannot. And the contemporary figure chronicled, too, can wound and be wounded in ways that Galileo or Lavoisier or Einstein cannot.

Writing history of science as it happens violates social contracts on several levels. The conversation between the past and present becomes in studies of contemporary science a conversation between historians and scientists (or others involved in contemporary events). The lone researcher immersed in archives becomes the interviewer and colleague, even the participant. This is a profound social transformation of the process of doing history and while many other disciplines have constructed ways of managing this social dimension, historians have not. Historians by disciplinary agreement do not (cannot) employ the methodological buffering mechanisms of the journalist, anthropologist, or sociologist. A historical account is specific rather than universalizing; actors in a historical account are so important they must be named; and the historical account depends on long-term, intimate knowledge, rather than the temporary expertise of the journalist who can move on after a day, week or month to another story.[13]

Social relationships though widely viewed as trivial are therefore central to doing contemporary history. By looking closely at them we can gain insight into that central problem of science studies: the status of knowledge as a social product. The knowledge we produce depends on social networks and when we do contemporary history those networks intersect in revealing ways.

I consider the social nature of historical knowledge by examining relationships first, between the historian and the subjects of oral histories; second, between the historian and those whose activities he or she chronicles (scientists, lab workers, administrators); and third, between the historian of contemporary science and other historians who work on persons who are (long) dead.

WHAT IS AN ORAL HISTORY?

I hesitate to raise the question because so many of my colleagues seem to have an answer that goes something like: "An oral history is a primary source that can be used to answer questions about what happened when, to whom, under what conditions". My answer is at odds with this one so I want to approach it obliquely. I consider two stories, the first about an oral history interview that I conducted, the second about an autobiographical fragment and its uses by biographers of Marie Curie. I mean to suggest that the oral history is like the autobiography, interesting primarily for the insight it provides into the informant's ways of understanding or making sense of the events in question. But this introduces into the process of conducting interviews a social barrier that the first definition elides. The differences between interviewer and subject become even more awkward for the two fundamentally do not agree about what is going on in the room. Some of those promoting the use of oral histories push this awkwardness even farther. They argue, for example, that the purpose of oral histories is not to obtain information but to flatter the aging scientist so that he or she will donate personal papers to a particular archive or provide the researcher with access to those papers. Others suggest that oral histories are a necessary concession to living scientists who will be offended if they are not interviewed. Are resources devoted to such interviews well-allocated?

My first story is about an interview during which I asked the subject a vague question along the lines of "How did that begin?" He provided a description of a particular day some 45 years earlier, recalling a meeting, who attended, who said what, and how the issue was resolved. I was at the time beginning to suspect that oral histories had limited value. Participants could sometimes shed light on general questions and it was often helpful to understand how they retrospectively viewed issues or questions. But my interview subjects rarely remembered details even when prompted with chronologies I prepared from archival records. And so when my subject presented so many specifics I was impressed.

Some weeks later I ran across a published memoir by this subject and found that it described precisely the same events that he had so successfully described in our discussion. I realized that my informant could recall these events because he had recently written them into a narrative: He had already done the work of making them into a story and he could produce that story as a response to any related question. The account he presented to me was a constructed story for which he had made decisions about what to exclude and what to emphasize. I was not dealing with an exceptionally fine memory. The work of retrieval was invisible in our interaction — had he remembered it all suddenly when he wrote the essay? Had he consulted records? Had there been confusion on any points? The existence of the text changed my assessment of his instant account and it changed my interpretation of the interview. When an individual provides testimony describing events in his or her own life what exactly are they providing to the historian?

My second story focuses on a minor detail included in every biography of Marie Curie that I have read. The detail in question is the frozen surface of water in a basin in Curie's garret room during her early years in Paris. Her biographers uniformly report that Marie Curie's brother-in-law, Casimir Dluski (married to her sister Bronya) referred to her early years in Paris as her 'heroic period', a time when she ate radishes for dinner, and had to break the ice in the morning to wash her hands. Most biographers cite Eve Curie's biography of her mother, *Madame Curie*, as the source for this detail but the original source is not Eve, or Dluski himself, or his wife, but Marie.[14]

The 'heroic period' comment, and the frozen water surface, are self-reported by Marie in her autobiographical sketch, written some twenty-five years after the events in question and published as an addendum to her biography of her dead husband:

> "The room I lived in was a garret, very cold in winter, for it was insufficiently heated by a small stove which often lacked coal. During a particularly rigorous winter, it was not unusual for the water to freeze in the basin at night; to be able to sleep I was obliged to pile all my clothes on the bed covers".[15]

None of her biographers seem to recognize that it matters who reports such things. One cannot take the comment at face value ("Casimir called it her 'heroic period'") but must see it as Marie's reconstruction of her early student days, a reconstruction in which she permitted the reader access to both her own physical trials and her brother-in-law's mockery. Robert Reid, in his 1974 biography, thinks he needs to point out that many students in the Latin Quarter lived in cold rooms, as though the issue were whether Curie's life had been more or less difficult than that of other students.[16] But the ice in the basin was a detail that Curie herself chose to emphasize her transformation as an aspiring member of the priesthood of science.

The cold garret room was used by Curie explicitly to suggest the force of the change in those early years. "It would be impossible to tell of all the good these years brought me", she wrote, and after describing her difficult physical circumstances added that "this life, painful from certain points of view, had, for all that, a real charm for me". It permitted her to devote herself to her studies, as "a new world opened to me, the world of science, which I was at last permitted to know in all liberty". Curie linked the rejection of her body and its needs to her apprenticeship as a scientist and the frozen water revealed what she wanted to show: The monasticism of science and her own personal awakening through which the constraints of the body (its need for food and warmth) became irrelevant.[17]

Both the oral history and the autobiography, then, permit us to understand how someone makes sense of their life, how they have chosen to tell the story, and how they connect events. To use the autobiography in this way is not a betrayal of trust for there is no implied trust between the historian and the subject who is long dead. In the oral history, there is indeed an implied trust.

THE SOCIAL OBLIGATIONS OF THE HISTORIAN

The second social circle I want to explore is the one that the historian enters when he or she chooses to observe in a laboratory or to work on topics in which many of those involved are still alive. The modern laboratory has become an important site for sociological and anthropological studies of science.[18] For the historian, too, it has a certain appeal and many laboratories have opened their doors to historians as observers or participants. In this setting scientists seem to expect that the historian's final text will provide an omniscient perspective from which it will be possible to see who was right or wrong, or who should be celebrated and who criticized. But in practice the historian in a laboratory is a social actor like everyone else and social alliances can form whether we seek them out or not. Race, gender, class or age can bind the historian to some persons in a laboratory in ways that may affect the collection of information. Personalities can matter as much as intellectual priorities. And if one is working in a lab where scientists disagree — which is to say virtually any lab — personal alliances can affect access and credibility.

There is furthermore the problem of what to include in one's text. Something that might appear neutral to the historian could be interpreted by competitors or grant reviewers as a sign that funding should not be renewed. Evidence that the research plans were flawed (or that they were the focus of a bitter internal debate) might be used against those observed. So I want to introduce here a set of questions that emphasize the power of historical narrative to affect people's lives. What are one's obligations to the people in a working laboratory? Is it fair to report that a person made a mistake that set the lab back? That someone is disliked by their colleagues? That someone is considered incompetent? Such assessments do affect lab dynamics, just as they affect any institutional group. Are they irrelevant? They are not to the historian. Mentor relationships matter; personal idiosyncracies matter. Would an imaginary historian working in T.H. Morgan's fly lab at Columbia University in the 1910s have been able to provide an account of the lab's work without mentioning its complicated social dynamics? Without mentioning H.J. Muller's contentious relationships with co-workers? And would an account that left out all the petty problems of managing that lab (problems that have been used with great insight to understand laboratory practice by Robert E. Kohler) be a historical account?[19]

Historians do have responsibilities to living participants, though I have struggled myself with how far those responsibilities extend. The Atomic Bomb Casualty Commission, the post-war biomedical agency I studied, employed for a time a person in a prominent position who was profoundly disliked, apparently for valid reasons. In the archives there are thick files documenting this person's offenses. In my book about the project, however, I did not mention these problems. My silence was motivated partly by the fact that he was alive; partly by the fact that his transgressions were not central to my concerns. Yet I knew that had I been present at the laboratory in Japan when he was in power his actions would have mattered a great deal to me and to all those whose work I sought to chronicle.

Some authors have managed to navigate these waters rather well. Robert W. Smith's account of the space telescope project includes some anonymous characters — technicians who are unidentified, and so on — but the big players, those who have the largest stake in these events, are identified and pinned down, their views specifically addressed, and, in some cases, their mistakes presented for the world to see. At some points the authors make apologies for those who were fired or reprimanded by presenting 'the other side of the story'. At the same time, they offer a frank treatment of serious problems and their text captures the difficulty and complexity of the project.[20] The final account is sympathetic and at the same time it is filled with names and details about who did what and how decisions were made. Their solution combines diplomatic point-counterpoint presentations of controversies with the occasional use of anonymity and the careful defense, as far as is possible, of those who are made to look bad by the story. These are strategies familiar to journalists but rarely used by historians.

Sharon Traweek, as insider-turned-historian-turned-anthropologist, developed a different solution. In her *Beamtimes and Lifetimes*, Traweek undertook to explain the culture of a laboratory in which she had at one time been an employee. She returned originally expecting to write a history of the laboratory but became interested instead in how stories about the lab's past were used by participants. Eventually her project became anthropological rather than historical. This shift had some advantages that Traweek does not explicitly invoke but that presumably made her work easier, for it shifted her attention from the specifics of individuals to the broader culture, from a historical account to an anthropological account in which individual identity is less at stake. Her characters are students and senior scientists rather than specific individuals.[21]

The sociologist or anthropologist can thus make this social dimension abstract by the use of pseudonyms and other camouflaging devices. But what can the historian do when the events reported could damage the career or life of a person? There are several options. The historian can leave out of his or her narrative anything that makes actors look bad, petty, wrong, mistaken or careless. One way to manage this is to shift the level of analysis either up (to policy, planning, funding) or down to bring it tightly into focus on a particular technical skill and attend only to the dynamics of making this process work. The historian can also use anonymity, even if insiders will know exactly who is involved and despite the fundamental conflict involved. This is a respectful choice but at odds with historical standards. 'Informed sources' guide journalists, 'mid-level managers' speak to anthropologists, but only specific people speak to historians (history is a field that appeals, I suspect, to people who like specifics).

The historian can also adopt the position that he or she has extremely limited obligations to the people whose activities are being chronicled. He or she can therefore report what they do or say even if it will make them look foolish or damage their subsequent careers. This is one variant of the journalistic position and it is reasonably practical if one is constantly dealing with new populations (as many journalists are), or with public figures who expect endless grief. But journalists do not treat their favored sources that way because they need to preserve the relationship. It is not necessarily wrong to produce a historical text that enrages the subjects involved but it is not conducive to the preservation of those subjects as sources.

RELATING CONTEMPORARY HISTORY WITH THE PAST

My third and last social circle is that linking the historian of contemporary events to other historians who work on earlier time periods. In their essay on the disciplinary history of immunology, Warwick Anderson, Myles Jackson and Barbara Rosenkranz suggest that the problems of writing the history of immunology are problems of boundaries.[22] I want to suggest that many of the problems they discern in recent accounts of immunology are a consequence of the social problems I have outlined here: The fact that participants are alive, have a stake in the history, have drawn the boundaries and sought to control the boundaries. And I want to suggest that under these conditions it may not be possible to write the kind of historical account they advocate. History of contemporary or very recent events looks different, relies on different kinds of information, and can reach different kinds of conclusions.

The same could be said, of course, of histories dealing with the late middle ages or those dealing with the late nineteenth century. Different eras have been preserved through different records and one's attitude toward records is shaped (we have to admit) by how many one has. But the methodological and analytical differences between medieval studies and those of the late nineteenth century are not problematic within the discipline of history itself. The differences are recognized as having scholarly legitimacy. The same is not true of accounts of history as it happens.

We know too well that history itself has a history, that for the Greeks and Romans, history did not exhibit overarching patterns but only "the inexorable effects of human passions, weaknesses and ambitions. [...] The movement toward the modern" — in which human participation in time centered on salvation and progress — came to define the direction of history only in the seventeenth and eighteenth centuries.[23]

The historian Paul Lacombe observed in 1900 that "time is nothing in itself. It is only an idea we have".[24] I want to propose that time might be a social barrier. The distance generally presumed to be accessible to historians is not temporal but social. To do history we need to get away from the people involved and when we do contemporary history we cannot do that because they are our most important resource; they are our archives; they are (partly) our audience. We are socially embedded and the critical distance we need is effectively excluded from historical practice by the very attribute that defines history, its emphasis on individual actions.

Many of those working on contemporary topics are not sufficiently self-conscious about these social dimensions and about the real risks on both sides of engaging in the explicitly historical study of contemporary science. It is possible to construct informative historical narratives about contemporary science and such narratives can have enduring value. But like all historical narratives they are constrained by the peculiarities of source materials. Living participants change the nature of the history that can be written. Historians interested in contemporary science need to recognize that their account can have an impact on the people and events they analyze.

In his essay on history and sociology, Fernand Braudel, French historian and the inventor of the *longue duree*, repeated the "amused reproach brought by philosophers" that historians "never quite know what sort of history it is they are writing".[25] Historians who work on contemporary events need to heed the philosophers; they need to recognize the peculiarities of the history they write, its limitations and constraints, and its shaping by social mechanisms that do not shape other narratives of the past. Not knowing what sort of history one is writing can be peculiarly dangerous when one tries to write 'history as it happens'.

NOTES

1 See Gross and Levitt 1994. It is unclear why their attack on science studies came when it did. Has the field become more visible? Are scientific institutions threatened by academic critiques?
2 A general overview of the boom in studies of post-war science and technology is in *Osiris*, vol. 7 (1992), edited by Arnold Thackray. See also Lindee, Speaker, and Thackray 1992. Important recent works on post-war science include Smith 1989, Leslie 1993, Geiger 1993, and Kay 1993.
3 One good example is Smith 1989.
4 Collingwood 1968; For a similar, earlier articulation of these ideas, see Croce 1941.
5 Febvre 1965, 437; cited in Gagnon 1982.
6 Hofstadter 1968.
7 Shapin 1994, 409–417.
8 Schiebinger 1989; see also Schiebinger 1993.
9 Schiebinger 1987; see also Keller 1985.
10 For more on ELSI and its impact on historians, see my essay review, Lindee 1994.
11 Nelkin and Lindee 1995.
12 Lindee 1994.
13 I speak from experience: I worked for a decade as a newspaper journalist on small and large dailies in Port Arthur and San Antonio, Texas.
14 Currie 1937. The description of the heroic period appears on pp. 107–118, the frozen water on p. 118.
15 Curie 1923, 107–171.
16 "She *was* poor, but so were most students" (italics in original) in Reid 1974, 48.
17 Curie 1923, 170–171.
18 See for example Latour and Woolgar 1979, Knorr Cetina 1981, Traweek 1988 and Lynch 1985.

19 See Kohler 1992.
20 Smith 1989.
21 Traweek 1988.
22 Anderson, Jackson, and Rosenkranz 1994.
23 See Appleby, Hunt, and Jacob 1994, 56–60.
24 Paul Lacombe, cited in Braudel 1980, 65.
25 Braudel 1980, 65.

BIBLIOGRAPHY

Anderson, Warwick, Myles Jackson, and Barbara Gutmann Rosencranz, "Toward an unnatural history of immunology", *Journal of the History of Biology*, vol. 27 (1994) 575–594.

Appleby, Joyce, Lynn Hunt, and Margaret Jacob, *Telling the Truth about History* (New York: W.W. Norton, 1994).

Braudel, Fernand, *On History* (Chicago: University of Chicago Press, 1980).

Collingwood, R.G., *The Idea of History* (Oxford: Clarendon Press, 1968).

Croce, Benedetto, *History as the Story of Liberty* (London: Allen and Unwin, 1941).

Curie, Eve, *Madame Curie* (Garden City, NY: Doubleday, 1937).

Curie, Marie, *Pierre Curie* (New York: MacMillan, 1923).

Febvre, Lucien, *Combats pour l'histoire* (Paris: Colin, 1965).

Gagnon, Serge, *Man and His Past: The Nature and Role of Historiography* (Montreal: Harvest House, 1982).

Geiger, Roger L., *Research and Relevant Knowledge: American Research Universities Since World War II* (New York: Oxford University Press, 1993).

Gross, Paul R. and Norman Levitt, *Higher Superstitition: The Academic Left and Its Quarrels with Science* (Baltimore: Johns Hopkins University Press, 1994).

Hofstadter, Richard, *The Progressive Historians: Turner, Beard and Parrington* (New York: Knopf, 1968).

Kay, Lily E., *The Molecular Vision of Life: Caltech, the Rockefeller Foundation and the Rise of the New Biology* (New York: Oxford University Press, 1993).

Keller, Evelyn Fox, *Reflections on Gender and Science* (New Haven: Yale University Press, 1985).

Knorr Cetina, Karin, *The Manufacture of Knowledge* (Oxford and New York: Pergamon Press, 1981).

Kohler, Robert, *Lords of the Fly: Drosophila Genetics and the Experimental Life* (Chicago: University of Chicago Press, 1994).

Latour, Bruno and Steve Woolgar, *Laboratory Life* (Beverly Hills: Sage Publications, 1979).

Leslie, Stuart, *The Cold War and American Science: The Military-Industrial Complex at MIT and Stanford* (New York: Columbia University Press, 1993).

Lindee, M. Susan, Susan Speaker, and Arnold Thackray, "Writing history while it happens", *Knowledge: Creation, Diffusion, Utilization*, vol. 13 (1992), 479–486.

Lindee, M. Susan, "The ELSI hypothesis", *Isis* vol. 85 (1994), 293–296.

Lindee, M. Susan, *Suffering Made Real: American Science and the Survivors at Hiroshima* (Chicago: University of Chicago Press, 1994).

Lynch, Michael, *Art and Artifact in Laboratory Science: A Study of Shop Work and Shop Talk in a Research Laboratory* (London and Boston: Routledge and Kegan Paul, 1985).

Nelkin, Dorothy and M. Susan Lindee, *The DNA Mystique: The Gene as a Cultural Icon* (New York: W.H. Freeman, 1995).

Reid, Robert, *Marie Curie* (New York: Dutton and Co., 1974).

Schiebinger, Londa, "Maria Winkelman at the Berlin Academy: a turning point for women in science?" *Isis*, vol. 78 (1987), 174–200.

Schiebinger, Londa, *The Mind Has No Sex? Women in the Origins of Modern Science* (Cambridge, MA: Harvard University Press, 1989).

Schiebinger, Londa, *Nature's Body: Gender in the Making of Modern Science* (Boston: Beacon Press, 1993).

Shapin, Steven, *A Social History of Truth: Civility and Science in Seventeenth Century England* (Chicago: University of Chicago Press, 1994).

Smith, Robert W., *The Space Telescope: A Study of NASA Science, Technology and Politics* (New York: Cambridge University Press, 1989).

Traweek, Sharon, *Beamtimes and Lifetimes: The World of High Energy Physicists* (Cambridge, MA: Harvard University Press, 1988).

CHAPTER 4

Using Interviews to Write the History of Science

Soraya de Chadarevian

INTRODUCTION: SCIENCE INTERVIEWED

When I embarked on my first research project on twentieth-century science, I did not plan to do interviews, at least not straight away. The project concerns the making of molecular biology in post-war Britain, focused on Cambridge.[1] This is a field of which protagonists, journalists and historians had already given various accounts. Instead of collecting more participants' views, I was determined first to find my way through published material and archival sources so as to get my own ideas about what mattered in my story.

A number of events, however, led to a fundamental change in my attitude. First I discovered how restricted access was to the relevant archives. The Medical Research Council had funded molecular biology (or what was going to be called molecular biology) since 1947, and it was in the MRC archives that I expected to find a wealth of material. However, not only was there the restriction of the 30-year-rule which put heavy constraints on my research, but much of the material of interest was either personal, and therefore closed until 50 years after the death of the scientists (most of whom are still alive), or destroyed, or lost. Most papers I could consult presented polished formulations of negotiations completed and decisions taken. Gaining access to the archive involved lengthy diplomatic negotiations with the keeper, who seemed to handle requests quite erratically.[2] I had seen some documents which, according to the MRC rules, I should not have been able to consult and there were some photographs in the four slim files of archival material still kept in the filing cabinet of the director's secretary of what is today the Laboratory of Molecular Biology in Cambridge. But the study of those documents had raised more questions regarding the history of the laboratory and, by implication, of a good part of the history of molecular biology in Cambridge than they could answer. And most of the relevant personal papers were still in the hands of the owners.

A further reason for changing my attitude to the use of interviews came from one of the protagonists of molecular biology in the 1950s and 1960s whom I met at a social occasion. He had been involved in the Steering Committee which had set up and backed the research project on the history of molecular biology in Britain. Quite inquisitively he asked me if I had started work on my project. Reflecting on his curiously insistent questioning, I took his underlying question to be: How does it come you have not talked to me yet?[3]

I embarked on my first interviews after taking advice from an African historian and an anthropologist who had some experience with oral history. I chose the core members of what was to become the Unit and later the Laboratory of Molecular Biology in Cambridge as my main interlocutors but spoke also to secretaries, technicians and to the laboratory steward. In fact it was the latter who joined the group as a young boy at the end of the 1940s who first showed me around in the laboratory and introduced me to people whom he thought could be helpful. I had a set of questions which I tried to cover in each interview. They referred to how the group was drawn together, to the research culture which was developed, to the Cambridge context, the international standing of the laboratory, the use of the term 'molecular biology' and, more specifically, to the life, career and research work of the particular person I was talking to. The interviews were usually confined to one to three sessions of a couple of hours. Extensive preparation on my part preceded each meeting. The time-limit of the interaction was mostly set by the interviewees themselves. The average time granted for an interview was one hour, but once a session was underway it often stretched to two hours.

Have I been converted to oral history? In the following I will reflect on my experiences with this tool. First, however, it seems important to clarify my use of the term 'oral history'.

Especially in the British tradition where oral history developed in the context of social history, the term is closely connected to the political claim of writing history from below.[4] This is not applicable to my case. I am using the term in the broader sense adopted for instance by Anthony Seldon and Joanna Pappworth in their *By Word of Mouth. 'Élite' Oral History*. They define oral history as "information transmitted orally, in a personal exchange, of a kind likely to be of historical or long-term value".[5] In this definition, the term 'oral history' refers to the interview itself. The same term is also used to refer to histories based on interviews. I will be concerned both with interviews and with the use of interviews in historical accounts.

Many of my interviews were 'elite' interviews, in the sense that the interviewees felt that they had contributed, to varying degrees, to establish what had become a dominant way of investigation in the life sciences.[6] In fact, a remarkable number of the early protagonists of molecular biology and thus of my interviewees were Nobel Laureates or Fellows of the Royal Society. The obvious problem of lending our ears to elite informants is that we give voice to those who already have a voice in history. The political (subversive) use of oral history as a research tool thus seems lost. But even elite interviews can be very useful in eliciting otherwise unrecorded information and new perspectives. Moreover, as the editors of a special issue of the *Journal of Contemporary Ethnography* on studying elites stressed: knowledge of the functioning of elites is as important to a critique of dominant power relations as knowledge of the disenfranchised.[7] Considering the reasons which induced me to plan and undertake my first interviews, I have implicitly pointed to some of the advantages, but also some of the challenges of using interviews to write the history of contemporary science. It is generally known that twentieth-century science and science administration produces an abundance of paper. Many historians even avoid entering the field for this very reason. But quite paradoxically those historians who do venture into the field of contemporary science are confronted with an acute scarcity of written sources. Much of what used to be put on paper is not put on paper any more. Many archives are closed and many personal papers are still in private hands. In this situation, interviews often serve to guide us through the flood of published papers and can lead to the disclosure of personal papers; discussions with the actors provide background knowledge about decision making, networks, practices and routines which are usually not recorded and enrich our understanding of personal relationships and individual viewpoints.[8]

However, using interviews also presents a series of specific challenges to the historian. Doing oral history involves practical problems of interviewing techniques, transcriptions, limits of time and money, copyright, etc. There is an expanding literature especially in the social sciences which covers most of these issues. In addition each project involving oral history presents specific problems depending on where the funding comes from, on the social position of the interviewees and interviewer, on the topicality of and the interests involved in the questions at issue.[9] In the following I will touch on some particular problems relating to my project. My focus, however, will be on more general historiographical questions raised by the use of interviews. Drawing on my own experience and relating to current debates in the literature I will in particular examine the status of oral history as a historical source, its relation to written sources and the uses we make of it for the histories we write.[10] In connection with these issues I will further discuss the question of the relationship between scientists and historians and their respective accounts of the past.

None of these questions is new. However, oral history is not yet fully accepted as a 'serious' historical tool in scholarly circles.[11] And a glance through the literature quickly reveals that even historians who use oral history hold controversial views on these issues. As a growing number of historians move into twentieth-century history, and as scientists increasingly recognize the market for their memories, it seems crucial to develop a clear understanding of the possible uses of interviews in reconstructing the past and of the relations of the scientists and historians involved. Several of these questions confront the historian of contemporary technoscience, whether using interviews or not. I will therefore argue for taking up the challenges presented by the use of interviews and for making space for it among the tools of the profession.

THE HISTORY OF A TOOL

To put the following discussion on interviews as historical documents and on the relation of written and oral sources into context it seems useful to start with a brief history of oral history.

Paul Thompson, who pioneered oral history in Britain in the 1960s, has reminded us that oral history is both a new but also a very old form of history.[12] In non-literate societies history was transmitted orally. This tradition is still alive in the 'griots' of African villages, who can recite from memory the genealogies of families, the dynasties of chiefs and the succession of natural and political events up to ten generations into the past. Jan Vansina and other historians started collecting and analysing these oral sources in the 1950s and have thus contributed to the re-evaluation of oral testimonies as historical documents.[13]

In Europe, even after the advent of literacy, oral sources and witness accounts were consistently viewed as reliable sources of information. From Herodotus to Voltaire, all Western histories relied on such sources for their histories. Significantly, there was no gulf between the study of early and contemporary history, or between history and the social sciences. These divisions and the disciplinary differentiation that went with them, occurred only in the second half of the nineteenth century.

In the nineteenth century, a new kind of history, originating in Germany, based on the critical use of sources that had been developed in the field of classical philology, established itself in European universities. The most influential exponent of this new historiographical movement was Leopold von Ranke (1795–1886), professor of history at the University at Berlin. Concerned with the study of large-scale political events like the formation of states, nations and legal systems, history had to (and could) be exclusively based on the use of archival documentary sources. As the historical manuals of the time put it: no documents, no history. The social sciences were divided from this new history and developed their own distinct disciplinary methodologies: anthropology was proceeded by field study, sociology employed the interview and statistical analysis, etc. Oral sources in history continued to be used in political biographies, in local history and labour history, but rarely would they be treated as genuine documents, i.e. they would not be footnoted. Political biographies, in fact, often occupy the borderline between history and literature. Today, footnoting of oral sources is one important strategy in the current re-evaluation of oral history.[14]

Thompson has attributed the revival of oral history both to a change in historical interest and to a change in the media of communication. A rising interest in social history and in aspects of life which are much less well represented in archival sources like everyday practices and social patterns and relationships, together with an increasing interest in contemporary history promoted the use of oral history interviews as a valuable source. This change of interest contributed to a loosening of disciplinary divisions between (some) historians, sociologists and anthropologists and led to a cross-fertilization of methods.[15] A similar process occurred in the history of science in the last twenty years.[16]

The second aspect mentioned by Thompson regards the impact of the development of communication technologies on the kind of history we write. The tape recorder has, in many ways, done for the spoken word what the printing press has done for the written text. Telephone, radio, television and video have further contributed to the formation of what Gwyn Prins has described as a "post-literate [...] electronically oral and visual culture".[17] In recent political affairs like Iran-gate, recorded oral information proved more reliable and compromising than the written records. As Prins has argued, the same developments have also deflated the professional self-esteem of traditional document-driven historiography. According to Thompson, the recorded interview provides the historian with a double form of entry into this new situation: both as a tool for gathering evidence, and also as a new way of conveying history through the new media.[18] Many oral historians, therefore, argue for the production of high-quality oral documents which would be accessible to other researchers and could also be used in broadcasting and education more generally.

The technical capability of producing sound records was used to establish the first oral history programs at American institutions in the late 1940s. Projects in the history of science and technology followed programs in cultural, political and intellectual history.[19] In a recent essay, Ronald Doel has pointed out how the available collections, even if reflecting more traditional approaches to the history of science, can be read in new ways and mined to reconstruct unexplored and otherwise unrecorded dimensions of scientific activity, like the role and acquisition of tacit knowledge or the formation of informal networks.[20] He suggests various ways in which future oral history programs can be constructed to exploit still better these possibilities and argues for the particular importance of oral history programs for an era like the Cold War in which much of the science actually pursued remains hidden in the written records.

In the following I will be exclusively concerned with questions regarding the collection and use of research interviews. Doing interviews presents a series of challenges posed by the direct interaction with the interviewee which are absent when using interviews already deposited in archives. That many interviewers hesitate to deposit their interviews in archives indicates this difference. Yet even in the case of deposited tapes and transcripts it is important to clarify the status of interviews as historical sources.

RECOLLECTIONS OR HISTORICAL SOURCES?

How can we and why should we use interviews as a historical tool if — as Bill Provine has so succinctly summed up the objection to oral history — interviewees do not tell us what happened but give us only recollections of what happened and unfailingly reconstruct their stories in a way that will put themselves in a positive light?[21] Provine's formulation raises three interrelated issues which all bear on the question of the status of oral history as a historical source: the question of memory, the narrative structure of oral histories, and the specificity and value of accounts based on oral history in relation to traditional (written) sources. I shall deal with them in this order.

Experimental psychology has confirmed what we all know from personal experience: memory is much more a reconstruction than a reproduction of past events. It is not just a storage system of past impressions, but rather reconciles the past with the requirements of the present.[22] This means that memory is not only very selective, but often creative and continuously changing. Experience in interviews quickly reveals that memories are particularly unreliable concerning dates and the exact sequence of events. Routine patterns or strong personal experiences are more reliably remembered.

While little is known about how memory actually works, different theoretical constructs have been developed to account for the functions memory serves. Psychoanalytic theory has elaborated the notion of the personal myth by which individuals explain and justify themselves. According to this theory, cognition and memory are accommodated to serve this myth. Glen has argued that scientists not only form personal but also collective myths which structure group identities and serve social and disciplinary functions.[23] Hoddeson uses Vansina's distinction of 'mask' or public image and 'face' to explain and work against the 'distortions of memory'. According to Vansina, people of many cultures tend to construct two portraits of themselves. One is a mask or a public image which reflects social roles, statuses and moral principles. Behind the mask is the face which reveals more personal emotions and contradictory experiences.[24] According to Hoddeson it is the historian's task to dislodge the mask and to aim at a "deeper history".[25] I will come back to this view below.

Not only past and present, personal and collective interests but also interviewing itself molds the memories of the interviewee. McMahan has insisted that oral history interviews do not call up recorded information but represent an interpretative communicative experience in which interviewer and interviewee co-produce an account of the past. According to McMahan, the historian is "the catalyst for as well as a participant in the creation of the historical record".[26] The way questions are phrased and answers interpreted by the historian as well as the presentation of documents of the past can restructure the memories of the interviewee. Gender, class, age and the professional relationship of interviewer and interviewee further shape the outcome of the interview.

A further, not necessarily contradictory, perspective is to look at the narrative structures of oral history interviews. Most of the responses by the interviewee are embedded in a narration, presenting a temporal and spatial order of events.

According to McMahan, storytelling in interviews fulfills a double function: it provides a way to talk about the past by offering meaningful answers to the interviewer's questions. Again then, the interview is revealed as interactive and its product as a collaborative effort.[27]

The narrative structure of discursive praxis and historical writing has been critically analysed by Hayden White.[28] He has argued that narration endows accounts of the past with a content which is a function of the form. On this basis he has posed the question of the adequacy of narrative discourse in historical writing in contrast, for instance, to the chronicle. White's analysis presents a powerful tool to think about our own work as historians. It can also be usefully applied to investigate the narrative structures of scientists' accounts of the past and the relations between scientists' and historians' accounts.

Scientists have different ways of framing their stories. Much of what interviewees tell is autobiographical and therefore embedded in accounts which make sense and do justice to their own lives and aspirations. Speaking about their institutional and disciplinary affiliations, scientists have to position themselves carefully. Evaluations of colleagues or employers, especially negative ones, are frequently made only off the record.

'History' is often reduced to a collection of anecdotes which may be recalled over and over again, accompanied by the same amused reactions. Like photographs which dominate the memory of past events, these accounts get stabilized, while other events and circumstances fall into oblivion. According to an alternative presentation which often coexists with the first, scientific developments are made to follow a rational order, even if some 'detours' are taken into account.

A particular form of scientists' account of their laboratory practice is the scientific paper. A retrospective reconstruction of experiments and their results, framed according to standardized conventions, it is well known that the scientific paper does not do justice to the way research work normally proceeds. And yet, as Glen points out, the account given in scientific publications often becomes the source for what is told in oral history interviews.[29] Of this and the other constructions every interview gives plenty of evidence.[30]

If oral history accounts are so fraught by personal and collective myths and narrative structures, how can they usefully serve as historical sources? My exposition so far about the working of memory and the formation of narratives seems fully to confirm objections against oral history. But I would like to invite a different interpretation. Whenever I have encountered or recognized a particular pattern according to which a single interviewee or a group of interviewees has structured his or their account, I have found these patterns informative. These constructions, if not taken at face value but confronted with other evidences, can illuminate for example the position of an individual in a group, or the ways by which a group constructed its identity. Often, they can also be helpful in the interpretation of written documents. In this sense I disagree with some of the conclusions drawn by the authors cited above. Glen sees collective myths "as systems of thought in the service of the group rather than as historical constructs". He concludes that "perfect congruence and convergence in multiple accounts of the same event should perhaps invite scrutiny rather than satisfaction".[31] While I do agree that congruence of multiple accounts points to collective constructions, I would consider those as 'historical' in their own right.

Similarly, Hoddeson acknowledges that the "masks" which are encountered in every interview are a "crucial object of the historian's study". They provide a view of "what is real for our interviewees". Yet we have seen that Hoddeson argues that the historian must aim to lift the mask. This she can do by confronting the interviewee with contradictory evidence and demonstrating that the mask is a construction.[32] Again, I find Hoddeson's notion of a "successive approximation" to a "deeper history", i.e., one which does justice to all available evidence and thus represents the end of historical research, as inherently problematic.[33] In contrast to Glen I have carefully avoided talking about "distortions" or "aberrations" or "the dangers of narrative"[34] in scientists' accounts and would see the difference in historians' and scientists' accounts not in the degree of "distortion" but in the kind of questions that are asked. This does not mean that it is not useful to probe scientists' accounts, confronting them with other views and documents and trying to convince them to abandon their preferred stories. But what we will get will be new memories or a restructured story that tries to make sense of the evidences presented. If stories are the form in which people account for past events, we cannot escape their structure.

One of my most important gains from oral history, and a positive lesson from the different narratives I have received, has been to become aware of the multiple perspectives that constitute what seems to be a straightforward sequence of events, like for instance the founding of the Laboratory of Molecular Biology in Cambridge. The challenge then becomes to account for the plurality of stories while writing one story.[35] Steve Woolgar, following an ethnographical approach to science studies, has also pleaded that historians take a reflexive stance, especially with respect to the moral order of representations.[36]

Throughout I have assumed that oral sources are used in conjunction with written sources. Written sources will corroborate oral information, but the reverse can also be true. Seldon and Pappworth have pointed out that many 'troubles' encountered in interviews, like inaccuracy of information, personal biases, lack of perspective or superficiality, which are all commonly cited as shortcomings of oral history, are shared by some well respected written sources like letters or diaries.[37] They argue that oral evidence is "of the same status as an abstracted letter or a diary written on the same day as the events being discussed. It is all history in the sense that it describes, reports or assesses what happens, and it has already progressed along the road of selection and evaluation".[38] Like all these sources, the evidence of oral history has to be critically assessed and evaluated. Interviews offer the advantage that, as in a courtroom, seeing someone answering questions, may give clues to the assessment of what is said that written statements do not provide. Hesitations, embarrassment and body language can all provide such clues.

Direct encounters with actors present particular challenges to the historian. I see a clear indication of this in the fact that every time I told fellow historians who worked on earlier periods that I had talked to one of my actors, immediately and unsolicited they took a strong stance. Either they wished they could do the same with their long deceased figures or they were quick to state that they were more than happy that the people they were concerned with were all dead. What exactly do these reactions mean? What difference does knowing one's historical actors make?

On the one hand, oral historians agree that knowing own's actors adds important dimensions to historical accounts. Interviews can "add a great deal of color" and "provide a human dimension like a soul".[39] Historians can "gain a 'feel' " for the interviewee which often would not be apparent from reading what she has written. It becomes also "much easier to read between the lines if one has known some of the participants, their mannerisms, habits of thought, and, not least, the way in which their colleagues are likely to have regarded them". This will result "in a more responsible, vivid and truthful account".[40] For the case of very extensive interviews, Thomas Söderqvist has even spoken about the opening of an existential understanding of the interviewee and his life project.[41]

On the other hand, the documentation and interpretation of the full range of human expressions displayed in oral history interviews still represents a major challenge to historians trained in the study of texts. McMahan has warned her readers of using merely the written record of interviews which represents only " 'the cadaver of speech' " for interpretation.[42] The tone of the voice, expressions, body movements, and gestures all carry messages. According to McMahan, video-recording, even if intruding on the interview, would be a far better record. Glen has predicted that the video-record will "grow in value as our ability to interpret the whole range of human expression improves in the future".[43] Some of the more interesting 'problems' of oral history would thus stem from the complex and rich experience of the interviewing process which we are not trained to decipher or to use for the stories we write.

WHO SPEAKS FOR THE PAST?

Among the historians who use interviews to write the history of science there is no agreement in how they should interact with the scientists in constructing their historical accounts. This emerges clearly from the different conceptualizations of the interaction between interviewer and interviewee to which I have referred so far and from the different use of interviews in historical accounts.[44]

McMahan sees the interview as a communicative process in which scientists and historians cooperate in producing a new historical document. She distinguishes between interviews in which the historian acts in the single role of information elicitor and those in which he acts in the double role of information elicitor and assessor. The two situations lead to different communicative interactions. Being exclusively interested in the interviewing process itself, McMahan does not address the question, how the interviews are integrated into the final account produced by the historian.

Hoddeson introduces the distinction between "passive" and "interactive" interviews depending on the degree to which the historian intervenes in the interview. She makes the case for interactive interviews in which the historian controls and tailors the content using questions based on his study of other documents. In her view, the historian has to engage with his interviewee in a confrontational way to achieve real collaboration in the construction of the historical account.[45]

Bill Provine, in a talk on "Pitfalls and rewards of working with living scientists in writing the history of modern biology", suggested that historian and scientist should collaborate closely in the attempt to reconstruct passed events.[46] Provine was obviously reflecting on his experience in writing Sewall Wright's biography, for which he engaged in several hundred hours of interviews with his protagonist. While the tapes and transcripts are all deposited for the use of other historians at the Library of the American Philosophical Society in Philadelphia, interestingly enough, they are not footnoted in the published text. This seems to indicate that, for Provine, the result of extensive interviews is that scientist and historian can speak with one and the same voice about the past.[47]

Quite in contrast to these positions, at a course on oral history I attended the advice given was to intervene as little as possible in the narrative account of the informant. This involved making as little statement as possible about one's own person, up to becoming a non-person.[48]

Quite obviously the approach chosen varies with the aim of the historian and his view of what science is about. The technique suggested in the oral history course was quite explicitly directed at the reconstruction of 'life stories'. This is not the aim of 'directed' or 'research interviews' as I for instance pursue them. I am not so much interested in the life stories of my interviewees as in their participation and interpretation of particular events. Quite obviously, the distinction between these two uses of oral history is not clear cut, but I do think that some kind of distinction has to be made. In agreement with other positions presented here I have found it quite impossible not to make any statement about myself. To achieve co-operation I have always found it crucial to disclose the scope of the interview and to share at least some information about the background of my project and about myself.[49] I have found it particularly important to make clear that I felt competent to follow even intricate technical details of the scientific questions at stake.

A particular problem relating to my project[50] has been that some of my interviewees felt that the history of molecular biology had already been written by James Watson, one of the main protagonists of the field, in *The Double Helix*, by the historian Robert Olby in *The Path to the Double Helix* and by the science journalist turned historian Horace Judson in *The Eighth Day of Creation*.[51] Both Olby and Judson had already interviewed some of my interviewees extensively. In these cases it was important to explain how my research agenda was a different one.[52] Appreciating this fact, however, does not imply that scientist and historian have to share the same agenda. I will turn to this point in a moment.

Provine's approach aims at reconstructing discovery processes and the formation of new theories. The implication seems to be that there is one best (objective) account of these events which is supported by written sources but which scientist and historian can best reconstruct in close collaboration.[53] Hoddeson, at least regarding the reconstruction of 'technical histories' like the one of the first atomic bomb, seems to share a similar position.[54]

In a dispute over her recent *The Molecular Vision of Life*, Lily Kay has taken a diametrically opposite stance.[55] In answer to an attack by Robert Sinsheimer on her interpretation of the wider social agenda of the Rockefeller Foundation's funding for research at Caltech in the 1930s, Kay asks provocatively: "What is history of science and who should speak for the past?"[56] Kay has no doubt. The aim of historical scholarship is to demonstrate that science is a genuine historical process shaped by and shaping social and political agendas. The practicing scientist has no privileged access to this history. According to Kay, the fact that such an exercise is deemed to be subversive by scientists, underscores the essential tension between the two professions. Kay's response offers a firm answer to a question which was also raised in the workshop that gave rise to this volume: Is there a qualitative difference between scientists' history and historians' history? The answer here seems to me emphatically: yes. As we have seen, scientists' history is often reduced to a collection of anecdotes, or, as for instance in historical introductions to textbooks or also in personal accounts, presents a rational reconstruction of the development of scientific theories. In these accounts history proceeds by theoretical breakthroughs attributed to scientists of particular brilliance and insight. These histories often serve disciplinary needs like constructing a research tradition or legitimizing a new research field.[57]

However, I do not think it follows from this that scientists cannot be useful informants for the different kind of stories we want to write about recent science. First of all, it is an important task of the historian to spell out the uses to which history is put in scientists' accounts. Interviews are an important key to understanding the stories and anecdotes that circulate in a scientific community. Secondly, in all historical accounts the reconstruction of certain scientific events is important and scientists can contribute to it.[58] Finally, scientists are participants or witnesses of their particular time, of social and political events possibly not shared by the historian or lived from another perspective. And they can be usefully probed on that even if they may think that it has nothing to do with the history of their field.

I encountered a striking example of this with one of the core members of the group of researchers I am studying. During his career this researcher was involved in all kinds of extramural activities which would potentially make him one of my most useful informants. And his immense collection of papers was already minutely catalogued and deposited for consultation, subject to his permission. However, he had neatly divided his life into war work, research work, and governmental work and insisted that the first and the last were not pertinent to my project. I became convinced that this view was at least partly formed in response to the attitude of his scientific colleagues towards his science advisory and administrative activities. Only in the course of a number of interviews was his resistance to disclosing information regarding his not strictly scientific activities partially overcome.

RESPONSIBILITY, RIGHTS AND MONEY

From what I have said so far it will have become clear that there is an in-built tension in historians' interactions with scientists. On the one hand historians need the scientists' trust to elicit information from them. On the other hand, for interviews to be useful, historians should be able to use the information won in interviews in accounts which correspond to their own historical questions and professional standards. They will, however, always remain responsible to their informants.[59]

The tensions between scientists and historians can become even stronger if there are important issues at stake or if the historical research is supported by the actors themselves. The attack on Kay's book is an example of the first case: In her analysis of the new biology in the 1930s, Kay argues that the same interventionist approach which went together with the eugenic agenda of the Rockefeller Foundation also informs the politics of the Human Genome Project. This is an assertion which those who, like Sinsheimer, were among the prime movers of the Human Genome Project in which the prestige of a whole new research area and big money are at stake, can hardly leave unchallenged. Sinsheimer contested a specific interpretation of history, without having been one of Kay's informants. Otherwise the argument would have taken a different form. But the case makes clear that no twentieth-century historian can escape the risk of being challenged by the actors of the history she writes.

I have, in fact, become convinced that many reservations about using interviews are really arguments against writing contemporary history or the history of events the witnesses of which are still alive. I suspect that the same historians who do not accept interviews as historical sources would gladly use the transcript of an interview, or on the tape itself, if it dated from an earlier period. It is with this expectation that oral history archives are being built up.

Regarding the second reason for tensions between scientists and historians, Tatarewicz in this volume reminds us that a significant portion of historical work, at least in the U.S. is funded by the 'actors', i.e., by the same funding agencies which also supported the scientific or technological research or by professional organizations of the scientists.[60] By commissioning scholarly research, these agencies relinquish some of their control over history but may be repaid by gaining credibility. However as Tatarewicz points out, even under the best of conditions historians under such contracts must maintain the interest and support of their hosts. This raises the question of whether individual historians should be left to deal with the situation or if a more general solution could be found. Some historians and sociologists at the International Workshop on the Historiography of Contemporary History of Science, Technology, and Medicine, held in Göteborg in September 1994, were sufficiently worried about the independence of their work that the possibility of drawing up a charter which would protect their professional rights was seriously discussed. The possibility of paying scientists for their interviews and thereby acquiring the right to use the information freely was also raised, but rejected as 'cheque-book history'.

Scientists may well have preceded historians here. Videotapes of scientists interviewed by fellow scientists on their life and work are sold at prohibitive prices by commercial publishers.[61] If we agree that interviews with the actors are crucial sources for writing the history of twentieth-century technoscience, then these developments should trouble us.

ACKNOWLEDGEMENTS

Earlier versions of this paper were presented at the First Maastricht-Cambridge Wellcome Workshop on Problematics of the History of Modern Medicine and Biomedical Sciences at the University of Limburg at Maastricht, September 1993 and at the Historiography Group at Cambridge in February 1994. I thank the participants of these two meetings as well as the participants of the International Workshop on the Historiography of Contemporary Science, Technology, and Medicine in Göteborg in September 1994 for many useful comments. For suggestions regarding the revision of the paper I wish to thank in particular Thomas Söderqvist and Nicholas Hopwood. I further thank Ronald Doel, William Glen, Lillian Hoddeson, Frederic Holmes, Patricia Fernandez Kelly, William Provine and Thomas Söderqvist for making their contributions to the Working Conference on Interviews in Writing the History of Recent Science, Stanford, April 1994 available to me prior to publication. I gratefully acknowledge support of my research by the Wellcome Trust.

NOTES

1 See for example Watson 1968, Olby 1974, Judson 1979, Yoxen 1981, Perutz 1987, Crick 1989, and Abir Am 1992.

2 Access to the MRC Archives and its important holdings has been much improved in the meantime.

3 John Krige and Dominique Pestre who, however, worked in the institution of which they were writing the history, have stressed the importance of early interviewing for clarifying the relationships with the actors; cf. Hermann *et al.* 1990, 803–806.

4 See Thompson 1988, especially Ch. 1.

5 Seldon and Pappworth 1983, 4. The distinction is also drawn between 'oral history' in a more restricted sense and 'research interview'.

6 Seldon and Pappworth (1983, 6) define as 'elites' those persons "who rose to the top of their chosen occupation". For McMahan elite informants are persons who develop accounts of their involvement in controlling society as contrasted with non-elites who provide accounts of their lack of control (McMahan 1989, xiv and 30). As an important corrective for my account I have included among my interviewees researchers which resented and resisted the development of molocular biology or got unduly written out of that history.

7 Hertz and Imber 1993, 3–4. Also Sharpe (1991, 30), exploring the potentials and problems inherent in writing the history from below, comments that there is "no self-evident reason why the oral historian should not record the memories of duchesses, plutocrats and bishops as well as miners and factory workers".

8 For a more exhaustive list of advantages of oral history, see e.g., Seldon and Pappworth 1983, 36–52.

9 On the particular challenges faced by in-house historians see Tatarewicz (this volume).

10 In the present context I use the term 'source' in the general sense of material which forms the basis of any historical work. For a more specific use of the term source as indicative of a particular way of treating documents, see Wilson 1993, 301ff.

11 For recent objections against oral history that have appeared in print, see, e.g., Heilbron 1989 and Franklin 1994, 465, including Collins' response to Franklin's critique of his use of oral sources in Collins 1994, 497ff.

12 For the following see Thompson 1933, Ch. 2 and Thompson 1991.

13 See Vansina 1985.

14 Footnoting aims to treat oral sources on a par with written sources. However, some researchers also compare oral sources to the evidence anthropologists collect on their field trips and resist the idea of turning their tapes into public records.

15 In the last decades, with changing standards of objectivity, the interview had actually been discredited as a research method in the social sciences. A re-evaluation of qualitative research methods in anthropology and sociology rehabilitated the interview as a research tool also in these sciences; see Fernandez Kelly, forthcoming.

16 Ludmilla Jordanova, for instance, argues that only oral history can recover the complexity and intimacy of twentieth-century history of medicine ("Oral history and the history of medicine", talk presented at the Oral History Course at the University of Essex at Colchester, January 1993).

17 Prins 1991, 135.

18 Thompson 1991, 372 ff.

19 A list of the major oral history projects and repositories for the history of modern sciences can be found in Doel, forthcoming.

20 Doel, forthcoming. Doel's suggestion is a reminder that oral sources, like any other sources, can be read in different ways and do not represent an unambiguous record.

21 William B. Provine, "Pitfalls and rewards of working with living scientists in writing the history of modern biology", talk presented at the IS/HPSSB meeting at Brandeis University, July 1993.

22 For reference to the pertinent psychological literature and its relevance to a critical assessment of oral history discourse, see Glen, forthcoming.

23 Glen, forthcoming.

24 See Vansina 1985, 8.

25 Hoddeson, forthcoming.

26 See McMahan 1989, 5.

27 McMahan 1989, 80–96.

28 White 1987.

29 Glen, forthcoming.

30 Gilbert and Mulkay, analysing transcripts of interviews conducted with a group of biochemists, distinguish two different and concurring ways in which scientists report about the history of their field. According to their findings scientists often cast their accounts in empiricist terms wherein the driving force of scientific change is presented as deriving from the compelling force of 'key experiments'. Alternatively, they depict the past in terms of the interplay of contingencies and personal factors. Gilbert and Mulkay suggest that historians, instead of using scientists' accounts to write supposedly definite versions of what is going on in science, could with more advantage take up the goal of "describing and documenting the various repertoires and interpretative devices used by participants" on various occasions; see Gilbert and Mulkay 1984, 125. For a critique of Gilbert and Mulkay's program of discourse analysis, especially if presented as the exclusive goal of the historian, see Shapin 1984.

31 Glen, forthcoming. Obviously these 'systems of thought' can be informative in themselves (cf. below).

32 Hoddeson, forthcoming.
33 Hoddeson, forthcoming.
34 Glen, forthcoming.
35 For a similar conclusion see Tatarewicz (this volume). Also Provine (forthcoming) more recently concedes that "abundant evidence, far from enabling clear reconstruction of events and origins of ideas, often is consistent with plausible, different interpretations".
36 See Woolgar 1988.
37 Seldon and Pappworth 1983, 16.
38 Seldon and Pappworth 1983, 6. For a similar 'suspicion' of written sources, see Provine, forthcoming.
39 Hoddeson, forthcoming.
40 Cf. Seldon and Pappworth 1983, 127, 38 and 51.
41 Söderqvist 1995 and in this volume.
42 McMahan 1989, 107.
43 Glen, forthcoming.
44 As pointed out above we have to distinguish between the interaction of historian and scientist in the interview itself and the scientist's contribution in producing the historian's account. In the interviewing process a (new) historical document is created that the historian can use as such or in conjunction with other (oral or written) documents to construct his own historical account.
45 Hoddeson, forthcoming.
46 Cf. Provine, "Pitfalls and rewards" 1993 (*op. cit.*, cf. note 21).
47 Provine 1986. As already indicated, more recently Provine has adopted a more critical stance towards the use of oral history interviews as historical sources; see Provine, forthcoming.
48 Course on "Oral History and the History of Medicine" at the University of Essex, England, January 1993. The course was led by Paul Thompson with the assistance of Robert Perks, Curator of Oral History at the National Sound Archive in London.
49 This, after all, is also acknowledged by Thompson; see Thompson 1988, 199.
50 Regarding the importance of insider knowledge in the historian's interactions with scientists, see also Gaudillière in this volume.
51 Watson 1968, Olby 1974, and Judson 1979. Interestingly, judgments about which of these accounts was more reliable, faithful or just enjoyable varied widely. Watson's account, even if signalled as a "personal account", is often taken to give a 'true' picture. A curious position in this respect is taken by Rosalind Franklin's biographer, Anne Sayre, who takes Watson's memoir to be a faithful account of his perception of events at the time, but not as an account of what really happened. Her aim is to restore the truth about Rosalind's contribution to the discovery; see Sayre 1975, 191ff. For a historiographical critique of Olby's and Judson's approach to the history of molecular biology see Abir-Am 1985.

52 The task of writing the history of molecular biology was formulated and promoted on a first conference organized by a committee of the American Academy of Arts and Science which brought together historians, scientists and archivists already active in the field. From the beginning, the collaboration of scientists and historians and thus also the use of oral history were seen as a desideratum; see especially Benison 1970.

53 For an historical account also aimed at the reconstruction of a discovery process, but making a more critical use of oral history interviews cf. Holmes 1991–93. Holmes's interviews with Krebs are not only carefully footnoted, but are introduced like a second voice in the body of the account and clearly signalled as such by means of asterisks. For Holmes's own reflection on his experience with interviewing scientists see his contribution in this volume and Holmes 1984, 141ff and Holmes, forthcoming.

54 Cf. Hoddeson, forthcoming. In contrast to Provine's way to proceed, however, in the important book she co-authored all information gained through interviews is carefully footnoted; cf. Hoddeson 1993.

55 Kay 1993b.

56 Kay 1993b; see also Kay 1993a.

57 This does not imply that historians' accounts are free from disciplinary constraints or that there have not been scientists who have turned into sophisticated historians.

58 This aspect is supported even by some historians who otherwise hold a critical view about the usefulness of oral history interviews. According to Heilbron historians of recent science are dependent on scientists "not for recollections necessary to historical reconstruction but for instruction in science" (cf. Heilbron 1989).

59 Regarding the legal regulations of oral history interviews there is an interesting difference between the American and British ruling. While British regulations give the interviewee legal ownership of the copyright in their words, according to legal judgment in America the interview tape is a joint production of interviewer and interviewee. On the British regulations, see Ward 1990, Ch 3.

60 See Tatarewicz (this volume). The same situation applies to a certain extent to my project. The Wellcome Trust which funds my historical research was only marginally involved in the support of the group of molecular biologists I am investigating, but most of its funds do go to biomedical research. The support of twentieth-century history projects is based on the aim to promote a fruitful collaboration between practicing biomedical scientists and historians.

61 Life Science Communications (London) has started a video tape programme of scientific autobiographies. The first autobiography is of Sydney Brenner, a Cambridge molecular biologist, interviewed by Lewis Wolpert. Consisting of 15 videotapes and 180 pages of transcripts, it is sold at the pre-publication price of $1995 and will be sold on the market at $2995. My enquiries if the material will be deposited in the National Sound Archive of

the British Library or elsewhere for free use of researchers have so far remained without reply both from the producers and from the Archive.

BIBLIOGRAPHY

Abir-Am, Pnina, "Themes, genres and orders of legitimation in the consolidation of new scientific disciplines: deconstructing the historiography of molecular biology", *History of Science*, vol. 23 (1985), 73–117.

Abir-Am, Phina, "The politics of macromolecules: molecular biologists, biochemists, and rhetoric", *Osiris*, vol. 7 (1992), 210–237.

Benison, Saul, "On oral history: a personal view", pp. 96–113 in *Proceedings of the Conference on the History of Biochemistry and Molecular Biology* (Brookline: American Academy of Arts and Science, 1970).

Collins, Harry M., "A strong confirmation of the experimenters' regress", *Studies in History and Philosophy of Science*, vol. 25 (1994), 493–503.

Crick, Francis, *What Mad Pursuit: A Personal View of Scientific Discovery* (London: Weidenfeld & Nicholson, 1989).

Doel, Ronald E., "Oral history, archival interviews, and the historiography of modern science", in Horace F. Judson and Thomas Söderqvist (eds.), *Interviews in Writing the History of Recent Science* (Cambridge, MA: Harvard University Press, forthcoming).

Fernandez-Kelly, M. Patricia, "Making sense of the other: The evolution of the interview method in the social sciences, in Horace F. Judson and Thomas Söderqvist (eds.), *Interviews in Writing the History of Recent Science* (Cambridge, MA: Harvard University Press, forthcoming).

Franklin, Allan, "How to avoid the experimenters' regress", *Studies in History and Philosophy of Science*, vol. 25 (1994), 463–491.

Gilbert, G.G., Nigel and Michael Mulkay, "Experiments are the key: participants' histories and historians' histories of science", *Isis*, vol. 75 (1984), 105–125.

Glen, William, "The suspicious life of oral data and science history", in Horace F. Judson and Thomas Söderqvist (eds.), *Interviews in Writing the History of Recent Science* (Cambridge, MA: Harvard University Press, forthcoming).

Heilbron, John L., "An historian's interest in particle physics", pp. 47–56 in Laurie M. Brown, Max Dresden, and Lillian Hoddeson (eds.), *Pions to Quarks: History of Particle Physics in the 1950s* (Cambridge: Cambridge University Press, 1989).

Hermann, Armin *et al.*, *History of CERN. Vol. 2: Building and Running the Laboratory* (Amsterdam: North-Holland, 1990).

Hertz, Rosanna and Jonathan B. Imber, "Fieldwork in lite settings: introduction", *Journal of Contemporary Ethnography*, vol. 22 (1993), 3–6.

Hoddeson, Lillian, Paul W. Henrikson, Roger A. Meade, and Catherine Westfall, *Critical Assembly: A Technical History of Los Alamos During the Oppenheimer Years, 1943–1945* (New York: Cambridge University Press, 1993).

Hoddeson, Lillian, "The conflict of memories and documents: dilemmas and pragmatics of oral history", in Horace F. Judson and Thomas Söderqvist (eds.), *Interviews in Writing the History of Recent Science* (Cambridge, MA: Harvard University Press, forthcoming).

Holmes, Frederic L., "Lavoisier and Krebs: the individual scientist in the near and deeper past", *Isis*, vol. 75 (1984), 131–142.

Holmes, Frederic L.: *Hans Krebs. Volume 1: The Formation of a Scientific Life, 1900–1933. Volume 2: Architect of Intermediary Metabolism, 1933–1937* (New York: Oxford University Press, 1991–1993).

Holmes, Frederic L., "Scientific interviews and the working historian", in Horace F. Judson and Thomas Söderqvist (eds.), *Interviews in Writing the History of Recent Science* (Cambridge, MA: Harvard University Press, forthcoming).

Judson, Horace F., *The Eighth Day of Creation: The Makers of the Revolution in Biology* (New York: Simon and Schuster, 1979).

Judson, Horace F. and Thomas Söderqvist (eds.), *Interviews in Writing the History of Recent Science* (Cambridge, MA: Harvard University Press, forthcoming).

Kay, Lily, "Letters" (original title "An escape from history?"), *Engineering and Science*, (summer 1993b), 42–43.

Kay, Lily, "Caltech biology in perspective", *Science*, vol. 262 (1993a), 1995-1996.

McMahan, Eva M., *Elite Oral History Discourse: A Study of Cooperation and Coherence* (Tuscaloosa: University of Alabama Press, 1989).

Olby, Robert, *The Path to the Double Helix* (London: Macmillan, 1974).

Perutz, Max, "The birth of molecular biology", *New Scientist*, vol. 114 (May 21, 1987).

Prins, Gwyn, "Oral history", pp. 114–139 in Peter Burke (ed.), *New Perspectives on Historical Writing* (Cambridge: Polity Press, 1991).

Provine, William B., *Sewall Wright and Evolutionary Biology* (Chicago: University of Chicago Press, 1986).

Provine, William B., "Hazards and rewards of oral interviews in writing the history of modern biology", in Horace F. Judson and Thomas Söderqvist (eds.), *Interviews in Writing the History of Recent Science* (Cambridge, MA: Harvard University Press, forthcoming).

Sayre, Anne, *Rosalind Franklin and DNA* (New York: Norton, 1975).

Seldon, Anthony and Joanna Pappworth, *By Word of Mouth: 'Elite' Oral History* (London: Methuen, 1983).

Shapin, Steven, "Talking history: reflections on discourse analysis", *Isis*, vol. 75 (1984), 125-128.

Sharpe, Jim, "History from below", pp. 24–41 in Peter Burke (ed.), *New Perspectives on Historical Writing* (Cambridge: Polity Press, 1991).

Söderqvist, Thomas, "After the 200th hour: The A/effects of long-term interviewing for science biography", in Horace F. Judson and Thomas Söderqvist (eds.), *Interviews in Writing the History of Recent Science* (Cambridge, MA: Harvard University Press, forthcoming).

Thompson, Paul, *The Voice of the Past: Oral History* (2nd ed., Oxford: Oxford University Press, 1988).

Thompson, Paul, "Oral history and the history of medicine: a review", *Social History of Medicine*, vol. 4 (1991), 371–383.

Vansina, Jan, *Oral Tradition as History* (London: James Currey, 1985).

Ward, Alan, *Manual of Sound Archive Administration* (London: Gower, 1990).

Watson, James D., *The Double Helix: A Personal Account of the Discovery of the Structure of DNA* (London: Penguin, 1968).

White, Hayden, *The Content of the Form: Narrative Discourse and Historical Representation* (Baltimore: Johns Hopkins University Press, 1987).

Wilson, Adrian, "Foundations of an integrated historiography", pp. 293–335 in Adrian Wilson (ed.), *Rethinking Social History: English Society 1570–1920 and Its Interpretation* (Manchester: Manchester University Press, 1993).
Woolgar, Steve, *Science: The Very Idea* (Chichester: Ellis Horwood, 1988).
Yoxen, Edward, "Life as a productive force: capitalising the science and technology of molecular biology", pp. 66–122 in Les Levidow and Bob Young (eds.), *Science, Technology and the Labour Process: Marxist Studies*, vol. 1 (London: CSE Books, 1981).

CHAPTER 5

Writing the History of Space Science and Technology: Multiple Audiences with Divergent Goals and Standards

Joseph N. Tatarewicz

INTRODUCTION

For the most part, historians and others writing about past events, and actors, need only reach an accommodation between the evidence, their accounts, and their peers or other interested parties. In studies of very recent science and technology, however, actors and others with direct knowledge of the events in question can be a source of information that supplements and even corrects the usual sources. They can also shape, confirm, or contest the accounts developed by the historian. Actors' understanding of history is shaped by their own personal experiences, but no less so by their profession's commitments, goals, and its own historical accounts. The historians' understanding, is shaped by similar factors.[1]

This chapter addresses the issue of writing and presenting history for multiple audiences with divergent goals and standards, using a variety of documented interactions with the actors (correspondence, oral history, informal discussions) as well as published reviews of recent histories of big science and technology. It compares the assessments of the success of the works as seen by various parties and presented to various audiences.

Historians, particularly those of us working in very recent topics, happily and eagerly share much with sociology, political science, anthropology, and the other disciplines that contribute to the social studies of science and technology. I shall speak somewhat provincially about 'history' and 'historians', but in a much-expanded sense that includes these other approaches and those that fall somewhere in between. I shall use the distinction between 'historians' and 'actors' in a rather crude sense, simply to distinguish conveniently between those who study an activity and those who do that activity. In what follows I shall refer to events and incidents that stretch the conventions and standards for footnotes and documentation. In some instances my own notebooks and files are the only paper records supporting my assertions, and in some few instances only my own fallible memory.[2]

ACTOR-SUPPORTED HISTORY

Even before research begins, actors can influence the writing of history in a way unique to recent history: by providing the funding either directly by commissioning a history, or indirectly through a variety of mechanisms. Much of the history of recent science and technology — in the United States at least — is supported by the actors themselves through various mechanisms. Several large government agencies (National Aeronautics and Space Administration, Department of Energy, the military services) support history divisions that also let contracts to outside historians for studies of selected topics. Professional societies (e.g., the American Institute of Physics, Institute of Electrical and Electronic Engineers) support and house history centers. Various *ad hoc* history projects, aimed at documenting specific events or institutions have combined funding from the usual sources with actors' contributions.[3]

This basis of support enables the study of various scientific and technical developments and issues, and the historiographic issues that arise therein are dealt with elsewhere in this paper. But this enabling has another side — it means that with rare exceptions the historical work will be focused only on 'successful' or 'significant' projects, those so defined by the actors themselves and for which they are willing to provide the money. History offices or divisions within agencies and societies provide a most important inside locus of historical values, and often have limited discretionary funding to provide to the community for historically valuable studies. When the inside historians can decide what to commission, they have the discretion to follow their profession's values and even tackle histories for which there is no actor constituency. However, even under the best of conditions the inside historians must maintain the interest and support of their hosts. Under conditions of financial stress, they must be ever more entrepreneurial and raise funds for historical work from ongoing projects within their agencies. This means, at least, that only presently ongoing projects are available to be sponsors of their own histories. Maybe the Superconducting Supercollider will find a few dollars in its shutdown costs to analyze historically what happened, but that would be a rare exception. On the grand scale of initial choice of topics, our range of case studies is limited and partially defined by the actors' own interests and historical sense.[4]

The history of space technology and science in the United States illustrates this close tie between patron and historian. The history of spaceflight, like so many other technical fields, began with popular retrospective and promotional accounts by rocket engineers in the 1940s, continued by a popular journalistic history that appealed to an eager public in the 1950s. Only after Sputnik did government and academic historians, and a few social scientists, begin to address this history, stimulated by history programs at the National Aeronautics and Space Administration (NASA) in 1959, the military services, and the creation of a Smithsonian Institution museum devoted to air and space flight (NASM) in the 1960s. The 1980s saw more sophisticated historiographic approaches that tried to relate the technical and institutional histories to broader questions and approaches in general history, and that tried to break out of a preoccupation with a national (largely United States) focus. Throughout, there was a large but only partially accessible literature produced by the military services for their own use.[5]

The 1980s saw a number of developments in space history that reflected a maturing process. The role of Sputnik was reassessed and to some extent demythologized; a series of sophisticated studies went beyond institutional and project history; and spaceflight finally made inroads to mainstream history when Walter McDougall's *The Heavens and the Earth: A Political History of the Space Age* won the Pulitzer Prize for history.[6] The community gained coherence and solidarity through informal networks of NASA and NASM and a series of conferences held at Yale (1981), NASM (1982, 1987), and at NASA (1993).[7]

The pressures of the broader historical profession found expression in reviews of the literature that found much to praise, but also much to be done. Still needed were "broadly synthetic, contextual, and interdisciplinary studies [to] explore the meaning of [this] particular field of history in terms of what it means to others". Comparative studies were needed to counter the nationalistic orientation, and the field had to break loose from a preoccupation with NASA.[8] There were few studies of the aerospace industry, few of university space research activities, and few of other government agencies such as the Weather Bureau, Department of the Interior, and the military. These lacunae in part reflected the difficulty of gaining access to or using archival sources, and the success of the NASA history program in making easily available a wide range of records. The relative ease of working with NASA Headquarters sources, and the extreme difficulty of using records where history offices or archives were nonexistent, tended to focus effort. This was all the more important since in space history, as in other areas of recent technology and science, the historian works in veritable mountains of only partly organized papers, grappling with a vast and bureaucratic enterprise.[9] Orphaned or dead institutions and programs, generally with no benefactors, had to be addressed by other means. The quandary and challenge of the agency historian in lean times is to match the enthusiasm and resources of the actors with the agendas and standards of the historical profession and try to keep everybody happy.

I have worked on two substantial historical projects supported directly by the actors, as well as related interpretive exhibitions of historical artifacts. The Space Telescope History Project operated out of the National Air and Space Museum and Johns Hopkins University in the years 1982 to 1990 was supported initially by the very NASA office that built the space telescope itself, to which was later added a considerable infusion of funds and other resources from the housing institutions as well as the National Science Foundation. It was very costly and involved exhibitions of the Hubble Space Telescope hardware, the Infrared Astronomy Satellite, and other artifacts were supported by NASA and by aerospace industry firms. My current project, a history of planetary geoscience since Galileo, is supported by the NASA Planetary Geosciences Division within its Office of Space Science. All of these interpretive historical projects benefitted from the remarkable generosity and tolerance of their sponsors, all of them tested the limits of that tolerance, and all of them provided opportunities to probe the different values and agendas of historians and actors.

SPACE TELESCOPE HISTORY PROJECT

The Space Telescope History Project arose from the strongly felt awareness of the actors that theirs was an enterprise of historic significance. Conceived in 1980–1982 while the NASA History Office was between directors, the Space Telescope History Project was a direct arrangement between the NASA Astrophysics Division's Space Telescope Office and the partnership of the National Air and Space Museum and the Johns Hopkins University. With Paul Hanle and Robert Kargon as project co-directors, Robert Smith as the lead historian, and a host of part-time collaborators and graduate students, the project did the majority of its research 1982–1988. Out of it came a book, a series of articles, an archive, and a variety of exhibits and collected artifacts.[10]

To NASA's and the scientists' great credit, we were given an extremely long leash. Only a few passing incidents marred an otherwise productive relationship of mutual respect. In one instance, one of the telescope's scientists became initially agitated over our using his NASA records, which he had hoped to use himself to write a memoir. After an initially shaky start, he became one of our best allies. In another instance we had been granted access to high-level advisory board meetings that dealt with the telescope, with the proviso that we would not immediately or injudiciously reveal the details of discussions therein. In this we were basically following the Star Trek 'prime directive' of not interfering with the normal development of alien civilizations.[11] One of the NASA managers of our contract saw an opportunity to use us as spies, and attempted to debrief us about the meeting, explaining that was why he had us invited in the first place. When we refused to cooperate with this, there were a few veiled threats to our funding. This passed rather quickly, and the issue never came up again — at least not overtly. There were several mysterious instances where our funding was held up for various reasons, or where demands were made that we turn over various kinds of raw research materials or oral history interviews that had not been approved by the interviewees.[12] These were more troublesome, since one could never be sure whether the hold-up in funding was due to the normal bureaucratic complexities or something more nefarious. In these instances, we were able to use the Smithsonian and Johns Hopkins sponsorships as buffers, and financial contributions from other sources as shields.[13]

These incidents I do not regard as unusual, but rather to be expected. As historians, we were using a methodology and a set of standards somewhat foreign to that of the actors. To them, particularly since they were paying part of the bill, our activity appeared to be somewhere between consulting expertise and public affairs. On the other hand, they could have had complete editorial control by having a true 'court history' written by somebody to their specifications. This they chose not to do, mindful of the status of such histories. Thus when the actors commission a scholarly history, with credentialed historians, institutions, and presses, they relinquish some control but gain some credibility. Quite naturally, this issue of control and credibility is ever dynamic and constantly negotiated, reacting to a host of events and stresses that occur within and without the relationship.[14]

Except for these incidents, we were given unprecedented access to the developing project and its many meetings. We were encouraged to tell the story 'warts and all'. We were befriended by many project participants — from senior scientists to technicians — who fed us information, tipped us off, and in various ways offered and provided assistance. Often enough they wanted to be sure we were aware of their perspective, but this was useful and they were not in general pushy about it. This access continued even after the discovery of a flawed mirror, and even after Senator Barbara Mikulski waved the book at a NASA Administrator testifying before Congress. Reviews of the book in scientific journals have been overwhelmingly positive, and to my knowledge nobody has berated it for lack of accuracy because it was not written by professional scientists. True, the first edition of the book was published before launch, and thus dealt more with the engineering, management, and politics than with science. Even though it appeared long before the discovery of the main mirror's spherical aberration, there were still embarassments aplenty to be reported and discussed, and ample opportunity for NASA managers to fear its reportage.

In fact, our grace at the hands of NASA is all the more remarkable when one considers a recently published 'kiss and tell' memoir from deep within the space telescope project itself.[15] Eric Chaisson in his memoir provocatively entitled *The Hubble Wars: Astrophysics Meets Astropolitics in the Two Billion Dollar Struggle Over the Hubble Space Telescope*, recalls being told by a friend:

> "you are positioned as few others to make a contribution to the culture of the modern scientific endeavor, to burst the myth that science is lacking in human values. Only a rocket scientist actively plying the trade can inform the next generation of the true conduct of space science today".[16]

Chaisson was not himself a rocket scientist, but rather an extremely well-trained astronomer with a skill and penchant for popular writing. In charge of the Space Telescope Science Institute's public affairs and educational activities during the five or so years surrounding the launch and discovery of the mirror's flawed manufacture, Chaisson has candidly and ably chronicled those events from the perspective of his own personal involvement. His account nicely compliments our work, covering only partially overlapping chronological periods. In his account, not surprisingly, his own Space Telescope Science Institute appears as the White Knight, NASA as a hidebound and bungling bureaucracy, the industrial contractors as shameless profiteers, with the intelligence community on the sidelines watching 'the amateurs' with some considerable and security-classifed bemusement.

Chaisson's account is explicitly based on his own diaries, although there are a few in-text citations to other literature. A briefly annotated bibliography at the end of the book is all the reader has to go on for further investigation. Nonetheless, his account of events and his characterization of individuals and groups rings true, if at times self-serving and somewhat retributional. In instances where we had access to the same meetings and events, his reports are accurate. Chaisson's book has not been out long enough to see how it will fare with the scientific reviewers. I suspect it will take only a few lumps from historians with regard to its scant documentation. Whether its accounts will be contested by other scientists located elsewhere in Space Telescope will be interesting indeed to watch. My point is simply that Robert Smith and we, manifestly not ourselves rocket scientists, fared rather well with the actors and with our historian colleagues.[17]

This raises the issue of technical competence, and whether, to use Chaisson's phrase, it takes a rocket scientist to write about rocket science. Most recently and provocatively, this stance has been articulated in the book by Paul Gross and Norman Levitt, *Higher Superstition: The Academic Left and Its Quarrels With Science*. They write:

> "A serious investigation of the interplay of cultural and social factors with the workings of scientific research in a given field is an enterprise that requires patience, subtlety, erudition, and a knowledge of human nature. Above all, however, it requires an intimate appreciation of the science in question, of its inner logic and the store of data on which it relies, of its intellectual and experimental tools. In saying this we are plainly aware that we are setting very high standards for the successful pursuit of such work. We are saying, in effect, that a scholar devoted to a project of this kind must be, *inter alia*, a scientist of professional competence, or nearly so".[18]

This is another way of saying that the historian must understand the actors sufficiently to 'get the facts straight', and this is a requirement in history of science no less than politics, diplomacy, or any other area.

Few, if any historians of science would disagree with placing a high value on understanding one's subject. But the real quarrel often arises not in technical accuracy, but rather in the preponderance of emphasis of technical versus social and other factors. Here there can be genuinely defensible choices to be made and reasonable people, scientists and historians, can disagree. Scientists and some historians might prefer to spend most of the pages on the scientific and technical issues to the exclusion of the social and political, and others the reverse. It is a question of audience and purpose of the literary work one is producing. Indeed, this choice of audience and emphasis often enough divides historians from sociologists and other kindred scholars just as much as it divides us from the scientists and engineers.[19]

Gross and Levitt make a valuable point when they worry about commentators who rely on second and third-order generalizations or metaphorical treatments of science, especially in such formalistic areas as physics and cosmology. There is no substitute for detailed study of the science one is treating. Rather, it seems to me to be a question of time investment. And this means time investment not just in learning the science in question, but time spent in the narrative and analysis that one produces in order to bring the reader along. There is often more than enough science and technical content in any historical episode we characteristically treat to fill volumes, and we must condense or cut some of this in order to explain esoterica to the reader and to address other questions and issues.

This brutal zero-sum game is often disappointing to the scientist who would revel in greater scientific detail. As Harry Collins has noted, it is a matter of what questions and whose questions we use as a basis on which to interrogate the sources.[20] William McNeill has characterized the historian's craft in part as pattern recognition and has emphasized the important choices that shape the account:

> "Only by leaving things out, i.e., by relegating them to the status of background noise deserving only to be disregarded, can what matters most in a given situation become recognizable".

Often the dispute between historians and actors can be understood in terms of what one decides to relegate to background noise. The actors may understandably be disappointed to find an outsider relegating important events or personalities to secondary status. But the story or analysis presented must inevitably compete with the wealth of detail, and no account could include everything. McNeill notes that "exhausting the relevant documents before they exhaust the historian" is not a viable historiography, and Paul Forman has poignantly declared, "History is not one damn thing after another". Historians seem to have come to terms with the historicity of their own craft, and have largely abandoned the hope for a single, definitive, true account. Actors, however, can be decidedly uncomfortable with the notion of multiple and superseded histories. Many scientists have yet to come to terms with their own historicity and this gives, much to their chagrin, historians the last word. As Gabriel Motzgin has noted, "Memoirs were written with the idea of contributing to a future history, of influencing future historians, of winning the battle in history that had been lost in life".[21]

Besides the impossibility of having the last word, there can be other factors at work as actors and historians assess the success of a historical work. McNeill notes that:

> "sometimes an historian turns traitor to the group he studies by setting out to unmask its pretensions. Groups struggling toward self-consciousness and groups whose

accustomed status seems threatened are likely to demand (and get) vivid, simplified portraits of their admirable virtues and undeserved sufferings".

Historians can also sin just as grievously, in the eyes of the actors, by failing to emphasize enough positive and uplifting aspects of the area about which they write. With the end of the Cold War came reconsideration of basic tenets of government support of science and technology that had been in place since the Second World War. These irresistable currents, turbulent and unpredictable, swept big science projects of all kinds out of complacency and into anxiety. The cancellation of the Superconducting Supercollider, for example, sent the most severe shock waves through the physics community and other sciences as they struggled to understand what had happened and what were the implications. At the same time, the Space Station was only the most visible and controversial of large space projects to undergo extreme scrutiny and challenge with an uncertain future. The current news and "Washington Reports" sections of science newsletters and magazines are replete with accounts of these events, as well as intense soul-searching to understand their meaning and implications. These events and insecurities have made the actors in disciplines traditionally well-supported by the government extremely sensitive to historians' and commentators' accounts.[22]

Gross and Levitt go much farther than simply demanding technical competence and seem to argue, simply, that science requires special erudition but history or social science does not and may easily be picked up. Indeed, they make this most explicit: if the entire humanities department of MIT were to disappear, they say with some glee, the scientists could "with enough released time patch together a humanities curriculum to be taught by the scientists themselves" that would be "no worse than operative". Not only that, but the scientists might thereby prove that the humanities types were totally dispensable. This asymmetry of authority and expertise is a common enough claim, and has a long history in the struggle between the 'hard' and 'soft' sciences and the humanities. Calling the attempt to explain science as culturally constructed they call "futile and perverse", yet in virtually the same breath assert that this can explain the cultural constructivists themselves.[23]

Gross and Levitt's attack coincided with a number of other developments that pitted historians and other critical writers about science against some scientists who felt under attack. As the U.S. National Academy of Sciences and the National Science Teachers Association finished several years of developing national science standards for schools:

> "some scientists were up in arms over the description of the philosophy of science. Instead of saying that researchers make discoveries, the [draft standards] document described science as 'a social activity' of 'constructing knowledge', and emphasized the 'tentative nature of scientific knowledge'".

In a further draft, now under review, these offensive characterizations have disappeared replaced by bland language:

> "Many individuals have contributed to the traditions of science. Studying some of these individuals provides further understanding of scientific inquiry, science as a human endeavor, the nature of science, and the relationships between science and society".[24]

These differing assessments by actors and historians take place largely within a sheltered academic environment. Scholarly books, articles, reviews and the curricula at colleges and universities only occasionally reach broad public scrutiny. I think we can write for our professional historical audiences, and even write primarily for the actors, without selling our souls. It all gets negotiated in the gritty details, and I hope the anecdotes I recounted show that the actors are capable of both trying to shape the account in their own interests, as well as granting the historian considerable leeway even when it goes against their own interests. Indeed, my experiences have been reasonably good in that regard. In other, more public contexts, however, actors and historians grapple with similar issues but higher stakes. Because of the greater public scrutiny, these contexts can be less forgiving.

EXHIBITIONS AND ARTIFACTS

Exhibitions and artifacts are peculiar 'texts' to produce and explicate. However one considers the concept of a text, such displays are in some way statements and embody assertions and judgments. With fifty words or less and no footnotes, the literary medium is brutal. NASA and its aerospace contractors are of course more than eager to get their hardware and preferably their own message into the Smithsonian Museums. Concurrent with work on the space telescope history, I served as curator of several exhibitions of space science hardware — ranging from millimeter- square detectors to the structural prototype of the Hubble Space Telescope itself, the 43-foot tall Structural-Dynamic Test Vehicle. Any public display — especially a prestigious space like the National Air and Space Museum — is highly valuable and thus holds the potential for being highly contested. In most instances, mounting an exhibition of space artifacts means getting those artifacts donated and hopefully raising substantial financial contributions to support the exhibition as well. In the case of the space telescope prototype, which had spent five years in outside storage at the Lockheed Missiles and Space Company near San Francisco, it meant a lengthy and expensive restoration process as well. For this had been a prototype vehicle, a testbed, that had been built from leftover parts, endlessly tinkered with, and then disassembled and tossed on the scrap heap once it had served its purpose. The 'congealed' history contained within this artifact became the basis for disagreements on how to restore and exhibit the vehicle.

NASA saw the opportunity to have a life-size version of the Hubble Space Telescope, all dressed up in silver and gold foil, towering over the historic artifacts in Space Hall. Lockheed saw the opportunity to showcase its work just down the street from Capitol Hill. The superb museum restoration technicians saw the opportunity to recreate the Hubble Space Telescope in all its aesthetic glory, just as the original and only flight spacecraft had come out of the factory, like the mint-condition aircraft and flight-spare spacecraft exhibited throughout the museum.

As Robert Bud points out, museums are problematic locales for engaging in subtle teaching and exposition.[25] Aviation and space museums especially, with the grandeur and commanding presence of their artifacts, encourage peering "in wonder and awe at the machines" and gazing "at the scientific instrument with dimly comprehending respect". I wanted to restore and exhibit the space telescope prototype as itself, not dress it up for a costume ball. I wanted, as some horrified museum professionals put it, to put up a huge, ugly, grey, tin can right in the middle of Space Hall, one of the prime pieces of real estate, where it would surely cause adjoining property values to plummet. I had two main educational goals, each of them quite simple. The actual spacecraft is huge, the size of a railroad tank car, and the most frequent comment I have heard is "I never knew the true size of that beast until I saw it in person". I even heard this from scientists and engineers who had worked on the program for years and knew in their heads every dimension and specification. My second educational goal was to showcase its ugliness, to show that even the most sophisticated and high-tech apparatus often comes from humble, ugly beginnings. This was done using a series of explanatory panels and paired photographs that showed the flight spacecraft and the test vehicle going through their respective evolutions and assembly.

Once the opponents had gotten over their shock, and thanks to a window of opportunity provided by some fortuitous administrative confusion at the NASM, the artifact went up. NASA allowed Lockheed to donate the artifact (it was government property), Lockheed provided a large amount of cash, sent its engineers to Washington to assist, and various divisions of the museum swung enthusiastically behind the project. Over the course of two years, a host of wonderfully talented and generous people worked to restore what looked like a big pile of scrap to its circa-1976 configuration, when the overall design was frozen some fourteen years before launch.[26]

In the course of doing the restoration in suburban Maryland, transporting it to downtown Washington, and assembling it in place the museum staff recapitulated some of the manufacturing process of the actual spacecraft. The documentary and photographic research and many hours of oral history interviews with the engineers and designers captured the history of the artifact and its role in the overall program. Here is an artifactual bit of 'congealed history' which is immensely rich, as each time the artifact was passed from one Lockheed development group to another modification was piled upon modification. We have few, if any, reliable historiographic techniques for dealing with such non-textual sources. We were fortunate in that some of this research could be incorporated into the more traditional book and article products as well as explanatary briefings for the docents who would lead museum tours.[27]

PLANETARY ASTRONOMY AND PLANETARY GEOSCIENCE

Turning to another area of space history in which I have worked I wish to address two separate aspects of planetary science — first, my earlier work on the state and status of planetary astronomy in the space age, and second a current project in which I am engaged focused on the general history of planetary science since Galileo.

In the first, I examined the interactions of astronomers and planetary scientists on the eve of the space age. This study, like the space telescope history, was less concerned with the science being done than with the material and institutional basis for doing such science. I examined the development of a program of studying the planets from the ground that arose within NASA because of a conviction that astronomers would not divert their attentions from the then-current studies of stars and galaxies to do 'remedial' planetary astronomy of benefit to and complementary to the program of spacecraft exploration of the planets.[28] This work, originally my dissertation, was largely supported by the Smithsonian Institution and the usual personal graduate student means. I thus relied on NASA and the actors only for information and access, not sustenance. Nonetheless, I found myself smack in the middle of a contested history, whether astronomers had ever, or with good reason, abandoned the planets in favor of stars and galaxies. As I examined the attention paid to planets versus stars and galaxies over the prior half century or so, it became ever more difficult for me to separate 'fact' from 'stance'. Finally, I resolved to waffle firmly and examine only the effect of the actors' contention that the planets had been abandoned, and how that particular version of history was marshalled to achieve their ends.[29]

I found that proponents and discipline builders had been quite effective, and had used many available resources, including history, to build the case for NASA becoming involved in ground-based planetary astronomy. With their version of history as a context, they had used various federal agency interests, interpretations of events such as dramatic revisions of the then-understood Mars surface and atmospheric conditions, arguments of the greater economy of ground-based work, etc., to prod the space agency into becoming the biggest supporter of ground-based planetary research. Frankly, I could not determine which version of the history of planetary versus stellar and galactic astronomy was the 'true' one. Private reviewers and friendly critics of my work at conferences, and of my various manuscripts, however, were sometimes eager to persuade me of their version of the story. I was admonished by non-planetary astronomers that I had been suckered into the planetary people's contention of underdog status. Astronomers did not neglect the planets, but even if they did it was with good cause since planetary astronomy was a field that had exhausted its possibilities, merited only the small amount of resources it had garnered, and attracted second-raters. On the other hand, planetary proponents often found my characterizations too mild, and urged me to tell the true story of how they had suffered at the hands of astrophysical bigotry. Some of these more pointed critiques made their way into the reviews, albeit in a much subdued form.

One astrophysicist and former observatory director, in a review of my book, found it "flawed by his acceptance of the myth that professional astronomers [...] virtually ignored the astrophysical study of the planets". In a criticism that seemed to boomerang he noted that professional astronomers did indeed turn their instruments on the planets at first, but after a few observations "there was little more to do, in comparison with the universe of discoveries waiting to be made in the stars and nebulae".[30] Planetary scientists, on the other hand, found it to be a welcome confirmation of their self-identity by a real 'science historian', although they might question some specific points and emphases, and would have liked more detail.[31] At least one historian was moved to investigate this alleged neglect of the planets and produced a superb analysis of the chief institutions and agendas of planetary science before the space age.[32] There was enough ambiguity in my conclusions that the various camps could see their own reflections in it. Interpretative flexibility is a characteristic not just of artifacts, but of our own historical or critical accounts.

The final area that I wish to address is a current project which is to write a history of planetary geoscience since Galileo. Since this project is ongoing, I shall have relatively less to say about it. Nonetheless, it shows the same sorts of patterns that I have noted above in discussing my other related historical projects. First, it was conceived and sponsored by currently active scientists interested in and concerned about the fragility of their own history. Second, they wrote a sketch of their own version of the history as part of the request-for-proposals, a sketch that was somewhat anachronistic and reflected their own disciplinary biases. Third, they were receptive and agreeable to deep revisions of this sketch, and willing to forego their own conceptions to allow the historian latitude and professional freedom. Finally, I have been the beneficiary of much solicited and unsolicited advice and assistance from the community, which reflects a mixture of those eager to ensure I 'get it right', and those eager and willing to help in other ways.[33]

FINAL THOUGHTS: DIVERSITY AND TOLERANCE

Gross and Levitt have engendered a great deal of controversy by their pointed charges. I would like to turn one of their complaints on its head. They write:

> "By the same token, perspectivism is highly sympathetic to the claim that the heretofore disempowered have the right to have their own 'narratives', their own particular accounts of the world, taken as seriously as those of the standard culture, notwithstanding differences and outright contradictions".[34]

I would accord even scientists like Gross and Levitt the courtesy of having their own histories, and of taking them seriously. It makes no sense for historians and other critics to disempower scientists and engineers and appropriate exclusive rights to their histories. Conversely, it is important that historians do not dismiss the criticisms of the actors. The universe of possible evidence in histories of contemporary science and technology is far too rich and varied, far too complex and overwhelming in its quantity, to inspire supreme confidence in any single account or perspective. It would be a pity for each of us to withdraw to preaching to our respective choirs.[35]

ACKNOWLEDGMENTS

Portions of this work were supported at various times by the National Air and Space Museum, Smithsonian Institution, the National Aeronautics and Space Administration, the National Science Foundation, and other organizations whose generosity and tolerance is greatly appreciated. All opinions and conclusions herein are those of the author alone, and should not be attributed to any other source. Portions were developed for a talk at the Royal Institute of Technology, Stockholm, 13 September 1994. I am grateful to the organizers of the International Workshop on the Historiography of Contemporary Science, Technology, and Medicine held in Göteborg in September 1994, especially Thomas Söderqvist, to Svante Lindqvist, and to the participants for many valuable discussions. I am also grateful to my many colleagues in space history who are too numerous to mention individually, and who would probably want to remain anonymous anyway.

NOTES

1 As Jeff Hughes said in his oral presentation at the International Workshop on the Historiography of Contemporary Science, Technology, and Medicine, held in Göteborg, Sweden, September 1994: "we have scarcely begun to think about the problem of agency in the construction of our own historical accounts" (cf. Hughes, this volume).
2 Histories and other critical studies of science and technology that can stand alongside those done by historians have been done by actors. For example, see Tatarewicz 1994.
3 Warnow-Blewett 1992. Even where historians are not paid a salary for their historical work directly by the actors, they often rely on history centers or documentary collections sustained by the actors.
4 Of course, historical agencies such as the National Science Foundation provide some support for topics and projects primarily of historical interest. On government history offices see: Hewlett 1975, Hewlett 1978, and Holl 1985.
5 For an overview of this literature see Pisano and Lewis 1988.
6 McDougall 1986 and Bulkeley 1991.

7 The Yale conference was published as Roland 1985; reports on the other conferences appeared as Hirsh 1982, Mack and DeVorkin 1982; Hanle and Chamberlain 1981, and Collins and Fries 1991. For thoughtful assessments of the state of the literature see Dupree 1986 and Swenson 1993. Critiques of the National Air and Space Museum's exhibits and research directions can be found in McMahon 1981, Batzli 1990, and Roland 1993.

8 Hansen 1989, 643; see also Corn 1985 and Mack 1989.

9 Söderqvist (this volume).

10 Smith *et al.* 1993.

11 Aant Elzinga in the discussion after the presentation of this paper at the Göteborg workshop, remarked on the dangers of moral partisanship and the importance of not becoming an actor. This stance, which has its roots in the ethnographic tradition, has largely been adopted by historians working in contemporary issues. However, there are contexts in which historians advocate unabashedly becoming an actor: see Berkowitz 1986.

12 Various organizations have adopted standards and guidielines concerning the treatment of oral history interviews, and especially the issue of editorial control and intellectual property. Among others, see American Historical Association 1991; Neuenschwander 1985.

13 Historian Peter Novick worried intensely about inherent conflicts of values in such so-called "public history" projects. See Novick 1988 and Graham 1991.

14 Indeed, histories contracted for or supported by nonhistorical organizations lacking history or archives divisions are often paid or administratively managed out of the public affairs or educational divisions. The historian thus appears as just one of a number of such interpretive activities aimed at supporting the organization. On the many uses of history see Graham *et al.* 1983; on the quandaries of sponsored history see Karamanski 1990; for a remarkable expression of why actors should prefer independent historical approaches see Jefferson 1994.

15 We were, to use Jeff Hughes' phrase in his oral presentation at the Göteborg workshop, "invited tresspassers" who ran the risk of becoming guests turned "traitors" (cf. Hughes, this volume).

16 Chaisson 1994, ix.

17 Book reviews of *The Space Telescope* are to be found as follows (by historians or social scientists): Barke, Richard P., *Technology and Culture*, vol. 32 (1991), 637–639; Bromberg, Joan Lisa, *British Journal for the History of Science*, vol. 24 (1991), 118–119; Darius, Jon, *Annals of Science*, vol. 49 (1992), 93–95; Logsdon, John M., *Sky and Telescope*, vol. 79 (1990), 501–503; Mack, Pamela, *Physics Today*, vol. 43 (1990), 72; Neufeld, Michael J., *Journal of Interdisciplinary History*, vol. 22 (1992), 551–3; Waff, Craig B., *Journal for the History of Astronomy*, vol. 22 (1991), 250–252; Wheaton, Bruce R., *Science*, vol. 247 (1990), 1240–1241; and (by physical scientists), Anonymous, *The Economist*, vol. 315 (1990), 98; Herbert, Roy., *New Scientist*, vol. 126 (1990), 61; Marshall, Laurence A., *The Sciences*, vol. 30 (1990), 50; Smith,

Harlan J., *Isis*, vol. 82 (1990), 600–601; Smith, Sir Francis Graham, *Contemporary Physics*, vol. 31 (1990), 271–272.

18 Gross and Levitt 1994, 235. One scientist, an astrophysicist, told me in all seriousness, "any scientist can write the history of his own field — it's not that difficult — but nobody can write the history of a field unless they have had the equivalent of graduate training in that field". He did allow that one could understand and write the history of technological areas without advanced training, since these were more simplistic and penetrable than science.

19 Graham *et al.* 1983.

20 Collins in the discussion at the Göteborg workshop. As Söderqvist notes (this volume), the overwhelming number and scope of actors in contemporary science and technology is a severe problem for the historian trying to make sense of the enterprise. These actors, and their associated social apparatus of committees, task forces, etc., are also a severe problem in the exposition as well. Mark Washburn, reviewing a NASA-sponsored history of Apollo program science, lamented:

> "It's a truly horrifying tale, and Vincent Price should narrate the Talking Book version. [...] Compton lays out the bureaucratic jungle in excruciating detail, recounting the birth and growth of practically every box on the Apollo organizational chart. It reads like a case study on the spread of a virulent disease" (Washburn 1990).

Paul Forman provides an excellent set of questions and a discussion of the relationship between historians' and actors' history in Forman 1983 and 1991.

21 McNeill 1986, 8 and 2; Forman 1983, 825 (emphasis original); Motzkin 1994, 106.

22 McNeill 1986, 6.

23 Gross and Levitt 1994, 69 and 243. This assertion of professional non-parity is lampooned by the old joke told over and again by historians of medicine: the aging physician, remarking that in his retirement intends to take up history is countered by the historian who says that in her retirement she shall take up surgery.

24 Culotta 1994, 1649; Press Release 1994.

25 Robert Bud, "Reengineering commonsense: pencillin and technoscience", unpublished paper for the Göteborg workshop.

26 Dean Randall, quoted in Bud (see note 25).

27 Documentation of the acquisition and restoration of the artifact is contained in the artifact and exhibit files of the Department of Space History, National Air and Space Museum. The general issues of the proper role of museum exhibitions in aerospace are discussed extensively in: McMahon 1981, Batzli 1990, and Roland 1993. Thomas Hughes discusses artifacts as congealed

history in Hughes 1986, and the use of artifacts as historical evidence is treated in Lubar and Kingery 1993. The decision to remain faithful to the artifact's history had one unexpected salutory result. When astronauts were preparing for the first repair mission, they found this vehicle to be an important training aid, Tatarewicz (forthcoming).

28 Tatarewicz 1990.

29 For this waffling and my refusal to endorse a particular theoretical framework I was quite properly and gently taken to task. Bruce Hevly found a "lack of a clear thesis in the author's voice" (*Journal of American History*, vol. 78 (1991), 1152) and Karl Hufbauer wished for a "greater willingness to rely on theory for guidance" (*Technology and Culture*, vol. 33 {1992}, 646–647.) See also Chandra Mukerji, *American Journal of Sociology* vol. 97 (1991), 903–905. Reviews of *Space Technology and Planetary Astronomy* include (by historians or social scientists): Hirsh, Richard, *Science*, vol. 251 (1991), 957; Mukerji, Chandra, *American Journal of Sociology*, vol. 97 (1991), 903–905; Evans, David S., *Social Science Quarterly*, vol. 72 (1991), 864–865; Hevly, Bruce, *Journal of American History*, vol. 78 (1991), 1152; Hufbauer, Karl, *Technology and Culture*, vol. 33 (1992), 646–647; Waff, Craig, *Isis*, vol. 83 (1992), 690–691; Logsdon, John M., *Science, Technology, and Human Values*, vol. 18 (1993), 260–263; Mack, Pamela E., *Journal for the History of Astronomy*, vol. 24 (1993), 306–307; and (by physical scientists): Hughes, David W., *Annals of Science*, vol. 49 (1992), 292–293; Cruikshank, Dale P., *Icarus: Journal of Solar System Studies*, vol. 90 (1991), 334–335; Osterbrock, Donald E., *American Historical Review*, vol. 97 (1992), 316–317.

30 Osterbrock 1992 (see note 29).

31 Cruikshank 1991 (see note 29).

32 Doel 1990.

33 A rather unusual proliferation of contract histories arose at about the same time causing some confusion, even among the actors themselves. The NASA Headquarters Planetary Geoscience Division contracted with me for a history of planetary geoscience; shortly thereafter the NASA Headquarters Planetary Astronomy Division contracted with other historians for a history of planetary astronomy, apparently unaware of my book on the subject. The Caltech Jet Propulsion Laboratory, itself a NASA contract center, arranged with another historian for a history of planetary radar astronomy. While we welcome so many opportunities for historians, it is a curious commentary on the interdivisional communication at NASA and on the self-perceived identies of various kinds of planetary scientists.

34 Gross and Levitt 1994, 38.

35 I would not necessarily accord all such accounts equal belief, but rather fair consideration. Two recent conferences have been notable in gathering together historically-minded scientists and science-minded historians to discuss historiographic issues. One on the history of astronomy (at University

of Notre Dame, Indiana, June 1993) and another on the history of geology (at University of California at San Diego, March 1994). At both these conferences there were serious, and even at times heated differences in outlook. But they were characterized overwhelmingly by mutual respect and colleagial interaction.

BIBLIOGRAPHY

American Historical Association. Oral History Evaluation Guidelines (Washington, D.C.: American Historical Association, 1991).

Batzli, Samuel A., "From heroes to Hiroshima: the National Air and Space Museum adjusts its point of view", *Technology and Culture*, vol. 31 (1990), 830–837.

Berkowitz, Edward, "History and Public Policy", pp. 414–425 in Barbara J. Howe and Emory Kemp, eds., *Public History: An Introduction* (Malabar, FL: Krieger, 1986).

Bulkeley, Rip, *The Sputniks Crisis and Early United States Space Policy: A Critique of the Historiography of Space* (Bloomington: Indiana University Press, 1991.

Chaisson, Eric, *The Hubble Wars: Astrophysics Meets Astropolitics in the Two Billion Dollar Struggle Over the Hubble Space Telescope* (New York: Harper Collins, 1994).

Collins, Martin J. and Sylvia D. Fries (eds), *A Spacefaring Nation: Perspectives on American Space History and Policy* (Washington, D.C.: Smithsonian Institution Press, 1991).

Corn, Joseph J., [Review of Bilstein, Roger E. Flight in America: From the Wrights to the Astronauts (Baltimore: The Johns Hopkins University Press, 1984)], *Technology and Culture*, vol. 26 (1985), 871–873.

Culotta, Elizabeth, "Science standards near finish line", *Science,* vol. 265 (1994), 1648–1650.

Doel, Ronald E., *Unpacking a Myth: Interdisciplinary Research and the Growth of Solar System Astronomy, 1920–1958*. Ph.D. thesis, Princeton University, 1990.

Dupree, A. Hunter, "[Essay review] The history of the exploration of space: from official history to contributions to historical literature", *The Public Historian*, vol. 8 (1986), 121–128.

Forman, Paul, "Independence, not transcendence, for the historian of science", *Isis*, vol. 82 (1991), 71–86.

Forman, Paul, "[Book review] "A venture in writing history", *Science*, vol. 220 (1983), 824–827.

Graham, Loren, Wolf Lepenies, and Peter Weingart (eds.), *Functions and Uses of Disciplinary Histories* (Boston: Reidel, 1983).

Graham, Otis L. *et al.*, "Roundtable: 'the ideal of objectivity' and the profession of history", *The Public Historian*, vol. 13 (1991), 9–23.

Gross, Paul R. and Norman Levitt, *Higher Superstition: The Academic Left and Its Quarrels With Science* (Baltimore: Johns Hopkins University Press, 1994).

Hanle, Paul A. and Von del Chamberlain (eds.), *Space Science Comes of Age: Perspectives in the History of the Space Sciences* (Washington: Smithsonian Institution Press, 1981).

Hansen, James R., "[Essay review] Aviation history in the wider view", *Technology and Culture*, vol. 30 (1989), 643–656.

Hewlett, Richard G., "Government history: writing from the inside", pp. 7–16 in Frank B. Evans and Harold T. Pinket, eds., *Research in the Administration of Public Policy* (Washington, D.C.: Howard University Press, 1975).

Hewlett, Richard G., "The practice of history in the federal government", *The Public Historian*, vol. 1 (1978), 29–36.

Hirsh, Richard. "Proseminar on space history, 22 May 1981", *Isis*, vol. 73 (1982), 96–97.

Holl, Jack M., "The new Washington monument: history in the federal government", *The Public Historian*, vol. 7 (1985), 9–20.

Hughes, Thomas P., "The seamless web: technology, science, etcetera, etcetera", *Social Studies of Science*, vol. 16 (1986), 281–292.

Jefferson, Edward G., *The Value of History* (Philadelphia, PA: Chemical Heritage Foundation, 1994).

Karamanski, Theodore J., *Ethics and Public History: An Anthology*, (Malabar, FL: Krieger, 1990).

Lubar, Steven and David Kingery (eds.), "History from things: essays on material culture" (Washington, D.C.: Smithsonian Institution Press, 1993).

McDougall, Walter A., *The Heavens and the Earth: A Political History of the Space Age* (New York: Basic Books, 1985).

Mack, Pamela E., [Essay review] "Space history", *Technology and Culture*, vol. 30 (1989), 657–665.

Mack, Pamela E. and David H. DeVorkin, "Proseminar on space history: the National Air and Space Museum, May 22, 1981", *Technology and Culture*, vol. 23 (1982), 202–206.

McMahon, A. Michal, "The romance of technological progress: a critical review of the National Air and Space Museum", *Technology and Culture*, vol. 22 (1981), 281–296.

McNeill, William H., "Mythistory, or, truth, myth, history, and historians", *American Historical Review*, vol. 91 (1986), 1–10.

Motzkin, Gabriel. "Memoirs, memory, and historical experience", *Science in Context*, vol. 7 (1994), 103–119.

Neuenschwander, John A.N., *Oral History and the Law* (Denton, TX: Oral History Association, 1985).

Novick, Peter, *That Noble Dream: The "Objectivity Question" and the American Historical Profession* (New York: Cambridge University Press, 1988).

Pisano, Dominic and Cathleen S. Lewis (eds.), *Air and Space History: An Annotated Bibliography* (New York: Garland, 1988).

Roland, Alex, "Celebration or education? The goals of the U.S. National Air and Space Museum", *History and Technology*, vol. 10 (1993), 77–89.

Roland, Alex (ed.), *A Spacefaring People Perspectives on Early Space Flight. NASA SP-4405* (Washington, D.C.: GPO, 1985).

Smith, Robert W., Paul A. Hanle, Robert Kargon, and Joseph N. Tatarewicz, *The Space Telescope: A Study of NASA, Science, Technology, and Politics* (New York: Cambridge University Press, 1989).

Swenson, Lloyd S., [Review of Martin J. Collins and Sylvia D. Fries, (eds.), A Spacefaring Nation: Perspectives on American Space History and Policy, (Washington, D.C.: Smithsonian Institution Press, 1991)], *Technology and Culture*, vol. 34 (1993), 969–973.

Tatarewicz, Joseph N. "The Hubble Space Telescope repair mission", in Pamela E. Mack, ed., *NACA/NASA Collier Award Winners* (Washington, D.C.: GPO, forthcoming).

Tatarewicz, Joseph N. [Review of Don E. Wilhelms, To a Rocky Moon: A Geologist's History of Lunar Exploration (Tucson: University of Arizona Press, 1993)], *Earth Sciences History*, vol. 13 (1994), 67–69.

Tatarewicz, Joseph N., *Space Technology and Planetary Astronomy* (Bloomington: Indiana University Press, 1990).

Press Release: Draft Science Education Standards Envision Alternative Approach (Washington, D.C.: U. S. National Academy of Sciences, 1994).

Warnow-Blewett, Joan, "Documenting recent science: progress and needs", *Osiris*, vol. 7 (1992), 267–298.

Washburn, Mark. [Review of W. David Compton, Where No Man Has Gone Before (Washington, D.C.: NASA, 1989)], *Sky and Telescope*, vol. 79 (1990), 278.

CHAPTER 6

Participant Observation and the Study of Biomedical Sciences: Some Methodological Observations

Ilana Löwy

INTRODUCTION

This chapter reflects upon the observation of medical science in the making. Between 1986 and 1990 I observed the introduction of interleukin-2 (IL-2) and interleukin-activated cells into cancer treatment in a major French hospital which I will call the "Cancer Foundation". My study, partly based on participant observation, was a venture into mostly uncharted territory. Traditionally medical sociologists have observed activities in the clinics but have seldom been interested in the development and stabilization (or destabilization) of knowledge and practices. Historians of science, and more recently ethnographers of laboratories, have been interested in the validation of new scientific knowledge and new practices at the bench. Finally, 'laboratory studies' have often included the observation of biomedical laboratories, but have seldom been interested in hospitals or therapy.[1]

Only recently have students of medicine begun to follow more closely the development of knowledge and practices in the clinics.[2] Clinical trials are an official — although by no means unique — method of validation of such knowledge and practices and, should potentially at least, be a highly rewarding object of study.[3] In practice, I discovered, the observation of a clinical experiment rapidly becomes a tangle of theoretical, practical and emotional problems.

NAIVE OBSERVERS AND NATIVE OBSERVERS

There are, to put it in a nutshell, two approaches to the observation of scientific or medical research, the 'naive' and the 'native'. In their pioneering ethnological study of a laboratory, Bruno Latour and Steve Woolgar strongly advocate observation of science in the making by naive observers, complete strangers to the culture of modern laboratories:

> "The dangers of going native are particularly marked in the study of science, because as analysts we are inevitably caught up in the 'social science' tradition with explicit attempts to mimic natural science, and because of the currently widespread acceptance of the methods and achievements of science in the culture of which we are part".[4]

According to this view only a totally naive observer may avoid the pitfalls of non critical acceptance of the scientists' point of view. A complete stranger is able to perceive the alien aspects of the observed world, to question practices and beliefs seen as self-evident by the insiders, and therefore to shed new light on familiar activities. Harry Collins has advocated a radically different approach to the study of science: 'participant comprehension' based on the acquisition of 'native competence'. Drawing on his and Trevor Pinch's experience in studying paranormal phenomena, Collins proposes that the disadvantages of 'participant comprehension' are largely compensated by the advantages of directly experiencing the problems of the scientists they study.[5]

The choice of the subject of my investigation — immunotherapy of cancer — was at the same time the choice of a status of a 'non-naive' observer. I was trained as an immunologist, and before turning to science studies I worked in a cellular immunology laboratory, that is, in a scientific domain directly related to the Cancer Foundation IL-2 trial. My access to the trial, and the possibility of studying it from the point of view of a 'participant observer', were directly connected to my previous professional experience. I did not perceive this fact as a major problem. There is a long sociological tradition of 'ex-natives' turned observers of their previous culture. In the early days of the Chicago school of sociology, Robert E. Park trained members of specific social groups to study these groups. A probation officer later studied youthful delinquents, a social worker-turned-sociologist studied the ghetto, and an ex-hobo wrote a book on hobo culture.[6]

Another experience is closer to my own. Ludwik Fleck, today perceived as the pioneer of the sociology of scientific knowledge, was a bacteriologist and an immunologist, a specialist of the Wassermann reaction for the detection of syphilis, so eloquently described in his main theoretical work, *Genesis and Development of a Scientific Fact*.[7] His reflections on science were firmly rooted in his experience at the bench. Fleck presented his investigations as an attempt to develop a new approach for studying science, the 'comparative epistemology'. One may propose, however, that one of Fleck's most lasting achievements was the fine-grained description of biomedical investigations in the making. As Jan

Golinski puts it, "Historians and sociologists [...] might still profitably consider whether their vision of scientific practice can match the elegance and subtlety of Fleck's".[8] An insider's knowledge, Fleck's example shows, does not necessarily hamper one's ability for perceptive observation. It does not guarantee, however, that every 'inside observer' will be able to develop such fresh insights on his/her 'native culture' as Fleck did.

Observers, native or naive, may disturb the observed setting. Ethnographers and sociologists are familiar with this problem, and students of present science discover it. Discussing with actors, interviewing actors about their role, confronting them with documents and with their previous declarations, may change the way they perceive past events, but also influence their present actions. My observation of the IL-2 trial may have affected this trial. I like to believe that any such influence was very minimal if it existed at all, but I cannot prove it. In addition, some of the participants in the trial saw me as an ex-colleague with an unclear professional identity. My request to become a participant observer in the IL-2 experiment was seen by some of the Foundation's immunologists as reflecting a secret longing to return to the laboratory. They found it difficult to believe that an immunologist with an adequate training and secure professional position would wish to permanently exchange the solid ground of experimental science for fuzzy historical or sociological investigations.

The scientists at the Cancer Foundation were not sure how to classify a fellow researcher who shared with them expert knowledge and familiarity with the laboratory culture, but professed radically different goals. Sometimes I was perceived by other workers in the immunology laboratory as 'one of us', sometimes as an alien. My external/internal status affected my self-perception during the IL-2 trial. I oscillated between two frames of reference: the 'scientific' and the 'historical/sociological'. Occasionally I experienced a feeling of superiority, due to my supposedly broader, 'bilingual' and 'bicultural' point of view; at other times I felt myself a 'native of nowhere' — an inadequate immunologist and an awkward historian. 'Going native' is perhaps helpful in studying modern science, but investigators who observe scientists' activities still need to decide how 'native' should one go, and for how long.

NEGOTIATING ACCESS AND SECURING COOPERATION: SCIENTISTS AND STUDENTS OF SCIENCE

The main goal of my study was to investigate interactions between scientists and physicians during the introduction of new practices into the clinics. A previous investigation, centered on the role of tissue groups (the HLA system) in kidney transplantation, pointed to the role of loosely defined terms and practices (in that case, of the term 'histocompatibility' or the compatibility of tissues, and of practices aiming at measuring this compatibility) in facilitating interactions between transplanters and immunogeneticists. This was, however, a historical study, focused on events that happened fifteen to twenty years earlier. The development, in the meantime, of reconstructed and standardized 'discovery accounts' by the relevant professional communities, made my task more difficult. Thus, the actors who started by investigating compatibility of tissues in the context of pathology studies and later shifted to a genetic framework of explanation, recontructed their accounts of the early developments of histocompatibility studies in order to harmonize them with their present scientific outlook. When confronted with a written record, some researchers nuanced their stories, other stuck to the 'official' version of the historical event that led to the development of test of histocompatibility of human tissues.[9] Direct observation of medical science in the making, I assumed, will limit difficulties linked to an *a posteriori* reconstruction of events. It might help therefore to elucidate the role of loosely defined terms and practices in cooperation between clinicans and scientists.

The IL-2 trial at the Cancer Foundation was not a 'traditional' clinical trial of a new drug. Drugs are as a rule first tested in the test-tube and in animal models of disease (a pre-clinical stage), then in patients (a clinical stage). During routine trials of new therapies, contacts between the research laboratory and the clinics, if any, tend to be limited to an exchange of information about the expected and the observed effects of the tested molecule. In contrast, the IL-2 trial at the Cancer Foundation was strongly 'research-laden'. It closely associated research and therapy and therefore, I believed, was particularly suitable for the observation of interactions between the laboratory and the clinics. Some technical details are necessary to understand the reasons for this specificity of the IL-2 trial.

Interleukin-2 is a substance secreted by white blood cells, and credited with the ability to selectively stimulate cells active in immune reactions. In the early 1980s, genetic engineering technology made possible the cloning and mass production of this molecule, and led to a search for its medical applications. At first, specialists viewed IL-2 mainly as a potential treatment for deficient immune mechanisms. The interest in IL-2 as an anti-cancer therapy followed the interferon crusade: a massive, scientific and financial effort to adapt another molecule credited with the ability to stimulate cells of the immune system — interferon — to the cure of cancer. Around 1983, when it became clear that interferon would not become 'the penicillin of cancer', interleukin-2 came to the fore as the next candidate for attempts at immunotherapy of cancer with recombinant molecules (that is, with molecules normally present in the body, produced on an industrial scale by genetic engineering methods). In December 1985, a group of physicians and biologists at the National Institute of Health (US), led by Steven Rosenberg published an article announcing that regressions of previously incurable metastatic tumors were obtained in patients treated with high doses of interleukin-2 together with interleukin-activated white blood cells (Interleukin-Activated Leukocytes or LAK cells). The IL-2/LAK therapy was expensive, labor intensive, and it induced severe iatrogenic effects in the treated patients (high fever, digestive problems, cutaneous eruptions, oedema, cardiovascular complications, delirium); but, Rosenberg and his collaborators affirmed, it worked.[10] The widely publicized announcement of a new therapy for advanced cancer (it was given a prominent place in the evening news of all the major US TV stations, and was the subject of a leading article in *The Wall Street Journal*) started a wave of IL-2/LAK trials in the US and abroad.

The Cancer Foundation's IL-2 trial closely followed the publication of the first report on the new therapy. The moving force behind this trial was a group of immunologists who had recently arrived at the Foundation and who aspired to a better integration of their professional preoccupations with those of the clinicians. A clinical trial of interleukin-2 would, they believed, promote this goal. The IL-2/LAK treatment developed by Rosenberg and his collaborators required the combining of skills developed by immunologists — the activation of white blood cells by interleukin in the test-tube and then the quantification of anti-tumor activities of interleukin-activated cells — with practices of medical oncologists: the administration of a new therapy to cancer patients and the evaluation of their clinical progress. The introduction of the new therapy would therefore quasi-automatically forward closer relations between the immunology laboratory and the medical oncology ward.

Asking for permission to observe the IL-2/LAK trial, I explained that the main goal of my study was investigation of the interactions between clinical oncologists and immunologists, during negotiations on access to the trial. Unexpectedly (for me) the scientists and physicians who organized this trial did not seem to accept my explanations. They were puzzled by my 'self-evident' supposition that relations between immunologists and oncologists were problematic. I was in turn puzzled by their reaction: it was hardly a secret that scientists and clinicans at the Cancer Foundation collaborated rarely, if ever. Promoters of the IL-2 trial often affirmed that one of their aims was to improve cooperation between the research laboratories and clinical units at the Foundation. The difference between their point of view and mine, I understood later, stemmed from a radically different evaluation of the nature of relationships between clinicians and scientists. I viewed clinicans and scientists as belonging to distinct and at least partly incommensurable social worlds. Scientists and physicians are socialized in a different way, possess different skills, have different goals and different systems of reward. For me, such 'structural' incommensurability between scientists and physicians existed also in sites (such as leading North American teaching and research hospitals) in which clinicans and biologists cooperated frequently, and in which the same individual could occasionally occupy both professional roles.

Physicians and biologists at the Cancer Foundation were on the other hand socialized within 'science laden' domains (immunology, oncology) that stress the role of 'cutting edge' laboratory research in solving practical medical problems, and which do not accord an important place to life-style variables in affecting the patients trajectories.[11] Cancer specialists who work in leading research/teaching institutions tend to strongly adhere to the credo that the development of efficient ways to prevent/diagnose/cure human malignancies will stem from a better understanding of biochemical and genetic processes underlying the multiplication of normal and malignant cells. According to this view, a close collaboration between the experimental oncology laboratory and the cancer ward is not only a self-evident, 'normal' event, but also a highly desirable one. It is the most probable way to solve 'the cancer problem'. The belief that an increase in funding for research — in particular, for fundamental and pre-clinical research — is the most efficient contribution to alleviating the suffering of cancer patients, is shared by scientists who study cancer, research-oriented clinicans, powerful cancer charities and even patients' associations.[12]

The assumption that collaboration between cancer researchers and clinical oncologists (in contrast, for example, to collaboration between fundamental scientists and engineers) is in principle self-evident and non-problematic, leads to the attribution of the difficulties encountered by clinicans and researchers working together to local, contingent factors: inadequate institutional structures, poor management, the absence of leadership and idiosyncratic personal conflicts. My request to follow the interactions between physicians and scientists at the Cancer Foundation was consequently interpreted as a wish to find out 'what had gone wrong' and possibly to report on it. The latter supposition might have been particularly disturbing because the Foundation depends on charity money, and thus on a 'positive' public image. The suspicion that an external observer will be a 'debunker' and 'muckraker' was probably attenuated by the fact that I was a trained scientist myself, and (hopefully) also by some interest in science studies. Later, when I became a participant observer, it was also attenuated by the fact that I developed friendly relationships with other participants and by their getting used to seeing me around. However, each time a potentially disruptive incident occured — an obvious error made by one of the workers, a violent verbal interaction between several persons or an indiscrete remark concerning a colleague — heads automatically turned in my direction and there was uneasiness in the air. I was not very happy with my role as a disturbing intruder. On the other hand I was not sure that it is possible — or desirable — to study present science while steadily maintaining good relationships with all the scientists one is studying.

LOOKING FORWARD AND THINKING BACKWARD: PARTICIPANT OBSERVATION AND HINDSIGHT

My efforts to explain my research goals to the participants in the IL-2 trials were at best partly successful. I was able nevertheless to negotiate access to numerous trial-linked activities such as laboratory studies of interleukin activated cells, internal meetings and sessions with external collaborators. I attributed this practical success to a vague sympathy of the trial leaders for 'the humanities', to their difficulty in finding a convincing excuse to refuse my request (researchers are expected to help their colleagues), to a curiosity about my study, and to the belief that a qualified laboratory worker, even with unusual goals, will bring more profit than harm to the trial. All these reasons could indeed have played a role in securing my access to the trial.

Later, however, I arrived at the conclusion that the main reason for the initial acceptance of my request was the expectation that I would be a reliable witness to the trial's predicted success. The testimony of an independent professional observer, the trial's organizers expected, might help to present the IL-2 trial as an exemplary cooperation between scientists and clinicians, and forward their aim to extend such collaborations at the Cancer Foundation.[13] This expectation was expressed ironically (irony is a frequent way of dealing with tensions in the clinics). I was occasionally introduced to external visitors with a short speech which stated (approximately): "ours is indeed a very important study. Look, we even have our own historian. That person over there is present at all our meetings and writes down every single word we utter, so that nothing is lost for future generations".

My status as privileged witness of future success became, however, more complicated when the clinical experiment did not turned out as predicted. The promising results achieved in the first small-scale clinical trial of IL-2/LAK were not duplicated in subsequent, larger trials. Interleukin did induce in selected cases (usually one to four percent of the treated patients) long-term regressions of advanced, incurable tumors, but critics of the new therapy claimed that it was far too dangerous, unpleasant, expensive and inefficient for large-scale application in the clinics.[14]

Another setback, which affected more specifically the Cancer Foundation's IL-2 trial, was a growing doubt concerning the necessity of LAK cells in IL-2 therapy. Investigators who attempted to reproduce Rosenberg's studies arrived at the conclusion that the addition of LAK cells does not significantly improve the clinical results of IL-2 therapy, while it greatly increases its risks and its costs.[15] The decision (initially promoted by immunologists) to test IL-2 therapy at the Cancer Foundation was made when the initiators of this clinical trial believed that LAK cells were an indispensable part of the treatment. However, when the trial started two years later LAK cells were no longer viewed as essential and were deleted from the trial, partly under the pressure of IL-2 producers, interested in making interleukin therapy less labor-intensive, thus cheaper and easier to market. This was a major difference, because the administration of IL-2 alone, unlike the administration of a combination of IL-2 and LAK cells, does not call for collaboration between immunologists and medical oncologists. Indeed, in some US hospitals IL-2 therapy was conducted exclusively by medical oncologists who viewed interleukin as an additional anti-cancer drug. In contrast, immunologists and oncologists at the Cancer Foundation maintained their close cooperation even when it became obvious that the injection of LAK cells is not an essential element of interleukin therpay. The continuation of this cooperation was legitimated by the search for correlations between the activation of 'tumor-killing' cells in the blood of IL-2 treated patients (measured by immunologists) and the regression of their tumors (measured by oncologists). The description of such correlations, the organizers of the IL-2 trial explained, would allow for the pre-selection of patients responsive to IL-2 therapy (only about a quarter of the treated patients reacted to IL-2), and then the monitoring of the progress of their therapy. It would thus increase the efficiency of this therapy, and decrease its (human and material) costs.

Two years of intensive efforts of researchers at the Cancer Foundation did not lead, however, to the uncovering of regular correlations between the activation of specific subsets of white blood cells by IL-2 and the regression of tumors. Other approaches were necessary to explain the significant involvement of immunologists in this clinical trial. My study of the IL-2 trial was scheduled as an observation of a (relatively) straightforward interaction between two groups of professionals which needed to work together in order to reach a shared goal. It increasingly became an observation of approaches which consolidated the initially fragile relationships between oncologists and immunologists. These approaches included material techniques such as development of pre-clinical studies which employed biological materials from IL-2 treated patients; social techniques such as the organization of common working groups and joint meetings, common use of material resources (reagents, instruments), sharing of services of technical personnel; and finally discursive techniques such as joint writing of grant applications, research reports and scientific articles.

The IL-2 trial did fulfill some of its organizers' goals: clinicans and scientists learned to work together, the trial led to the introduction of new techniques and new scientific instruments to the Foundation, immunologists obtained access to patient-derived materials (blood, tissues) and were able to develop new topics of pre-clinical studies while clinicans got access to new methods of evaluation of the immune responses of their patients. These achievements were, however, different from the predicted results of this trial: the development of a successful anti-cancer therapy and a public demonstration of the immediate benefits, for cancer patients, of a cooperation between oncologists and immunologists. The presence of 'our historian' was seen as an asset in the early stages of the IL-2 trial when a straightforward triumph was expected, but became problematic when this clinical experiment turned out to be a much more complicated enterprise. In the later stages of the IL-2 trial its organizers successively reformulated their definitions of the goals of the trial. The adaptation of the (supposed) goals to the (existing) outcome helped them to continue to represent their endeavour as a success. At that stage, I have the distinct feeling that the presence of a 'professional memory keeper' who carefully recorded past assumptions and predictions and who, moreover, was qualified to understand technical debates, was viewed rather as a nuisance. The result was that my position became increasingly uneasy during the later part of the IL-2 trial. The trial's leaders were more reluctant to discuss the latest developments with me, and restriced my access to information (this restriction was partly counterbalanced by the fact that at that time I became friendly with some of the participants who were glad to share with me the latest news about the IL-2 trial).

Writing down my experiences with the Cancer Foundation, I became aware of an additional difficulty. My understanding of this trial — as presented in this chapter — is mostly a belated one, achieved in the trial's late stages. For example, at the end of my study I arrived at the conclusion that the organizers of the IL-2 trial were not aiming at a vague improvement of the relationships between the laboratory and the hospital, but had from the very beginning had precise organizational goals, such as a thorough reorganization of the clinical research department at the Cancer Foundation. Many of the activities which I initially attributed to different and non-connected motives, had aimed, I decided later, to fulfill these organizational goals.

'Perception shifts' — a new way of seeing the observed phenomena, a reorganization of the pieces of a puzzle in a different pattern — are often seen as important sources of innovation in an investigation, be it scientific or historical. They may lead to a reexamination of past evidence and may stimulate new activities. A scientist who has a fresh insight about a previous experiment can usually repeat this experiment, add different controls, and change some of the parameters. A historian who has a new insight about a document can often return to the archives (or, in the photocopy era, re-examine her files).[16] But it is not clear what a participant observer should do with her belated understanding of the observed phenomena. It is difficult to provide reliable account of events that should have been observed several years ago, and impossible to find out what the actors would have answered in a given situation if they had been asked the 'right' questions, meaning the questions that would be asked if the end of the story were known in advance.

WHERE (AND IF) CLINICAL EXPERIMENTS END: SETTING LIMITS TO INVESTIGATIONS OF AN 'OPEN' DOMAIN

Historians need to define what is the 'core' and what is the 'periphery' of their study, what is its 'object' and what its 'context'. Before I began my observations of the IL-2 trial I assumed that the definition of the scope of my investigation would be facilitated by my choice of subject. The observation of the IL-2 experiment was scheduled to be a study of a single institution — the Cancer Foundation — and of a collaboration between two laboratories (the immunology laboratory and a small hematology laboratory) and one hospital ward (a medical oncology ward). My observations, I believed, would be circumscribed in space and in time, and would therefore be self-restricted. This was, I rapidly discovered, wishful thinking. The IL-2 trial, I learned upon my arrival at the Cancer Foundation, was shaped by governmental policies: an initiative by the French government to encourage the local biotechnology industry (thus the testing of 'French' IL-2) and its efforts to promote close collaboration between government-sponsored research laboratories (such as the immunology laboratory at the Cancer Foundation) and private industry.

The verb 'shape' is employed here in a near-literal sense. The final form of this trial was elaborated in discussions between IL-2 trial leaders and officials in the health and industry ministries. Some of these discussions also included representatives of the industry.[17] In addition, there were numerous bilateral negotiations between the trial leaders and industrialists who provided the IL-2 and part of the funds for its testing.[18] I knew about these negotiations — everybody connected with the IL-2 trial did — but had no access to information about them, besides the occasional hints of participants in these debates which I was unable to verify. In France, unlike the US, there are no public hearings on policy issues, access to recent governmental documents is usually highly restricted, and industrialists seldom provide complete and uncensored accounts of their present activities. I could at best try to guess how essential the information that I missed was, hoping that it was not very important.

Even a (nearly) free access to relevant information would not eliminate the need to make decisions about the scope of a given inquiry. There were obvious practical limits on the quantity of the material which I could take into consideration, and on my ability to analyze this information. Sometimes I felt envious of my colleagues who studied other topics. The problem of access to sources, and of the definition of the scope of a given inquiry, is common to all historical investigations but its extent may vary. In some aspects at least, historians of present science may be less lucky than their colleagues who study a restricted corpus of documents such as medieval medical manuscripts. In addition, some scientific communities (such as high-energy physicians or paleontologists) are more isolated from other social groups and their work is less submitted to direct scrutiny by administrators, politicians and the lay public than the work of researchers who are looking for a 'cure for cancer'. Studies focalized on these groups may thus be more easily restricted to a 'maneagable' scope.[19]

Ideally, the decision of where and when to stop looking for new material should be well grounded in reasoned analysis of the subject. In fact, my decisions to extend the investigation of a given topic, or to stop a given line of inquiry, were sometimes conjectural, often partly arbitrary, and always subjective. Participant observation generates, by its very nature, subjective knowledge, and the aim of my study was to produce a 'thick description' of a single clinical experiment.[20] Such 'thick descriptions' do not pretend to be complete, and their 'thickness' necessarily reflects the personal choices of the observer. But students of science still have to decide how 'thick' their descriptions need to be, and where the borderline between a 'thick' description and a 'thin' one is.

CAN ONE GET RID OF PASTEUR'S MAD DOG? STEREOTYPED IMAGES OF CLINICAL RESEARCH

Clinical experimentation in oncology is a domain which has an unusually high visibility. The announcement of an innovation in the domain of, for example polymer chemistry is only infrequently diffused by the media, and information about new developments in computer science may or may not reach non-specialists. But when credible experts proclaim that they have developed a new treatment for cancer, this statement as a rule becomes important news. This visibility has concrete consequences. One is a higher level of public scrutiny (cancer researchers have more frequently been accused of fraud than for example dermatologists or geriatrists). Hence the importance of preventing a public presentation of experimentation in the cancer clinics as a failure. Another is the higher dependence on public representations of the scientists' activities in obtaining access to resources. Cancer researchers who become 'visible' for the lay public without losing their respectable image (those interviewed on 'serious' television and radio programs and whose work is reported in 'serious' newspapers) find it more easy to obtain space, funds, institutional recognition, and to attract qualified collaborators. Specialists in this domain are well aware of (and sometimes quite explicit about) the importance of good relationships with the media in forwarding their professional goals.

At this point the student of science may be tempted to strongly condemn medical scientists who arrange for themselves interviews in magazines, complete with glossy color photographs, and appearances on television programs, in order to inform the public about 'breakthroughs' in cancer therapy and at the same time advance their own agendas.[21] Biomedical scientists have a direct interest in presenting their work as leading to far-reaching practical consequences. The evaluation of medical innovations is often a long and complex process, and for practical reasons it is difficult to delay the attribution of professional rewards for such innovations until it is possible to reach a clear cut conclusion about their practical value. In the absence of resistance from the relevant professional community, compensations attributed to authors of well-publicized innovations — including innovations which are invalidated later — became as a rule irreversible upwards moves.

But is self-interest the only reason for the frequent announcement for example, of the development of promising anti-cancer therapies? Let's examine more closely the background and meaning of optimistic media reports. Reports on the 'search for a cure for cancer' are shaped by a more general tradition, that of stereotyped images of medicine in the contemporary Western culture. Medicine is usually perceived as a modern source of miracles ('white images'), occasionally — and symmetrically — as a source of abuse and evil ('black images'). These all-pervading, stereotyped representations of medical research, I propose, affect not only lay persons but professionals as well. Medical scientists, science policy makers, physicians, nurses, and health administrators have all been socialized with such representations. They have learned about great past epidemics, devastating maternal and infantile mortality, surgical operations without anaesthesia and without asepsis, suffering without painkillers — then about the discovery of vaccines, insulin, analgetics, antibiotics, and how these 'great medical discoveries' changed everybody's life (in Western countries at least). Images such as Pasteur and the mad dog or Fleming examining a dish with a green mold have become part of the Western cultural heritage no less than those of Napoleon at Waterloo or Churchill smoking a cigar.

Simplified images, such as the representation of biomedical researchers as driven exclusively by the search of truth and the wish to cure diseases, may be refuted at one level and fully accepted at another. The organizers of the IL-2 trial were aware of the fact that the optimistic presentation of the prospects of this therapy in their grant applications to cancer charities was a goal-oriented rhetorical device and, sometimes at least, they believed that their optimistic evaluation would became reality. Similarly, my impression is that enthusiastic appreciations of the future of 'immunotherapy' or 'biotherapy' of cancer made by the advocates of these new treatments in the media were often much closer to wishful thinking that to self-serving manipulation. Biomedical scientists are often aware of the fact that their studies are shaped by personal and collective interests, but at the same time they perceive their research as motivated by a combination of authentic curiosity, a disinterested drive to solve 'puzzles of nature', and a sincere wish to alleviate the suffering of the sick. Discursive devices, such as cynicism, which mockingly exaggerate the self-centered interests of medical researchers, help to bridge the gap between such contradictory beliefs. Cynical declarations allow one to acknowledge openly the existence of 'impure' motivations in medical research, but at the same time to present these motivations as non-essential, secondary, just a good subject for a joke.

The need for reconciliation of the contradictory representations of biomedical research (scientists as 'ordinary professionals' and as 'secular saints', respectively) is relatively recent. When medical science was still young, it was possible to view the search for power and the ability to control others as a respectable component of the activity of medical investigators. Students of past medicine may come across declarations such as the one made in 1886 by Dr. John Billings, the president of the American Medical Association:

> "This is the age of machinery, of exchanges, of corporations, for all these correspond to one and the same fundamental idea. Men make machines to do what the individual cannot do; and they make them not only of brass and iron, but of men, for such an obvious source of power for men who can master the combination is not likely to be overlooked. One result of such organization is seen in our encyclopedic works of medicine; another in great medical journals; another in associations which seek to wield political influence; another in the comparatively recent attempt at collective investigation of disease".[22]

Students of present medicine are less fortunate. They have to grapple with the effects of stereotyped images of scientists' and doctors' activities not only on medical scientists but on themselves too. During my study of the IL-2 trial at the Cancer Foundation I frequently oscillated between 'black' and 'white' perceptions of this trial. Occasionally I saw this trial as a basically admirable attempt to find a cure for advanced cancer made by sincerely concerned investigators. At other times I perceived it as a basically senseless enterprise which brought much more suffering than relief to the enrolled patients, but, in contrast, procured numerous professional benefits to its organizers. Sometimes a nagging voice repeated in my ear: "your study of the IL-2 trial is but a senseless academic exercise. Only one thing counts: this therapy may save the lives of numerous cancer patients. And if not this one, certainly the next one". On other days I caught myself feeling angry because the researchers at the Cancer Foundation were not always as perfect as I wished them to be. My long familiarity with biomedical research did not protect me from a tendency to idealize medical scientists unconsciously. My success in freeing myself from the stereotypic images of medical research was at best partial. It is not clear if it is at all possible to avoid/neutralize such stereotypic representations of scientific research, and given this difficulty is it perhaps less problematic — or at least safer — to select less emotionally loaded subjects of study, such as for example, crystallography or the production of agricultural machines.

USES OF A 'THIRD CULTURE': PRESENT SCIENCE AND THE HISTORIAN

In this chapter I have discussed a study of medical science in the making made by a historian. I would like to conclude with a short discussion of two questions: Why study a science in the making? and Why historians?

In 1963 C.P. Snow proposed a tentative answer to the question: Why study present science? Snow's well-known essay *The Two Cultures* pointed to a growing split between literary intellectuals and scientists and proposed that this dangerous split be overcome through mutual acquisition of a better understanding of the other group's culture. In a sequel, *Two Cultures: A Second Look*, Snow advocated, however, a different remedy for the growing divide between the 'two cultures'. This divide may be attenuated thanks to the recent development of a group of specialists able to understand the culture of scientists and that of traditional literary intellectuals, but who do not belong to either of these cultures: "I was slow to observe", Snow explained,

> "the development of what, in the terms of our formulae, is becoming something like a third culture. [...] Some social historians, as well as being on speaking terms with scientists, have been bound to turn their attention, to the literary intellectuals, or more exactly to some manifestation of literature at its extreme. Concepts such as 'organic community' or industrial society, or the scientific revolution are dealt with, under the illumination of the knowledge of the last ten years".[23]

The 'third culture', as seen by Snow, is not merely an intermediary between the two others, but is a distinct entity with a rationale, methodology and agenda of its own. Its task is not to popularize modern science or to judge it, but to investigate how science is made by human beings and what effects it has on human society.

Thirty years later such a definition may still be seen as an adequate general description of investigations by students of science who strive to describe and analyze the activities of scientists. If students of science were to become a 'third culture' which mediates between the literary and the scientific ones, their studies should not be read exclusively by their colleagues (that is, historical investigations by historians, anthropologists' research by anthropologists, etc.), but also by members of the other two cultures, that is by scientists and by literary intellectuals. These groups may, however, have very different criteria for evalation of accounts of present science than professional students of science. Such accounts may be appreciated by scientists because they display fine-grained understanding of scientific ideas, because they 'ring true', or because they lead the scientists to see their practice in a new light. They may be appreciated by literary intellectuals because they stimulate fruitful debates in the humanities or because they are well written. Students of present science who attempt to balance their tasks as members of a small, highly specialized professional community and their aspiration to serve larger social goals need to decide what — if any — weight should be given to appreciation of their work by members of other professional cultures.

The second question is: Why historians?, that is, why should historians study present science and in what ways (if any) accounts produced by historians differ from those generated by other specialists, such as sociologists, anthropologists, psychologists, linguists and journalists? I propose that while historians cannot pretend to possess an 'epistemological superiority' or an institutionalized capacity to produce accounts which contain a greater share of 'truth about science', accounts produced by (good) historians do differ from the ones generated by other professional groups. Their specificity stems from the consistent effort to place specific developments in science within a larger temporal framework. The observation of the IL-2 trial led me to an interest in topics such as the history of experimental oncology, tumor immunology, the use of animal models of cancer, clinical trials in general and clinical experimentation with anti-cancer drugs, the development of medical oncology as a distinct professional segment, biotechnology in the US and in France, and the history of Cancer Foundation. These historical investigations helped me to understand specific events during the IL-2 trial through the elucidation of their links with long term trends in oncology and experimental cancer research.

Historical investigations have shown, among other things, the persisting tension between the use of standardized animal models by cancer researchers and the highly individualized malignant diseases encountered by clinicans; the influence of the large-scale screening of anti-cancer drugs organized in the 1950s by the National Cancer Institute (US) in establishing fixed patterns for testing new anti-cancer therapies; the role of clinical trials in shaping the professional culture of medical oncology; the shift from exceptional to the quasi-routine use of experimental therapies in oncology.[24] The history of experimentation in cancer clinics may explain the stability and resilience of selected practices at the bench and the bedside. The history of the Cancer Foundation may illuminate the roots of present tensions between researchers and clinicans in this institution, and the attractiveness, for the Foundation's directors, of the proposal to restore the cooperation between these two professional groups. In addition, a historical perspective may facilitate the (partial) liberation of students of the present science from the pressure of ongoing events, from the influence of one's preconceived ideas and stereotypic representations, and from the consequences of tense relationships with the scientists that one studies — that is, from the myriad of conceptual and practical problems evoked in my study.

NOTES

1 For a review, see Löwy 1994.
2 E.g., Baszanger 1991; Dodier 1993; Berg and Mol (forthcoming).
3 The history of clinical trials and of their social uses was studied by Harry Marks, e.g., Marks 1987.
4 Latour and Woolgar 1979, 39.
5 Collins 1984.
6 Emerson 1983.
7 Fleck 1979.
8 Golinski 1990, 505.
9 Löwy 1990. See also Jean-Paul Gaudillière (in this volume).
10 Rosenberg *et al.* 1985.
11 Some domains of medical activity such as treatment of a chronic pain (Baszanger 1993) strongly rely on behaviorist approaches; others domains (e.g., cardiology) combine behaviorism and 'classic' medicine: patients suffering from heart disease receive drug therapy or surgical treatment but are also expected to develop an appropriate life style, seen as an important element in promoting cure. In contrast, mainstream oncologists do not think that changes in life style may have an important influence on the evolution of a malignant disease.
12 Fintor 1991.
13 Charles Bosk (1992) perceptively described the pitfalls of being an external observer in the clinics.
14 Moertel 1986; Rosenberg *et al.* 1980.
15 E.g., Negrier *et al.* 1989.
16 Some experiments, for example in high energy physics, cannot be repeated, some archives are no longer accessible, some documents are lost; but historians and natural scientists are frequently able to retrace their steps.
17 France is a highly centralized country and it is not too difficult for enterprising science managers to find their way to government officials.
18 IL-2 had no marketing permit at that time; clinical testing of this molecule could be conducted only in close co-operation with its producers.
19 For the description of a 'close' scientific community, see Traveek 1988.
20 Geertz 1973. Historians do not record 'historical facts' found 'out there' either, but actively select their material and then organize it into meaningful

entities, a process which, as Erwin Panofsky (1965) explained, is not very different from the selection and organisation of "facts of nature" by scientists. Panofsky, incidentally, viewed the construction of 'scientific facts' by scientists as a well-established phenomenon which, he proposed, should be extended to the construction of 'historical facts' too.

21 The utilization of the media as a resource by scientists is probably encouraged in France by the absence of a tradition of investigative journalism and by a parallel absence of serious criticism of science in the media.

22 Billings 1886.

23 Snow 1963, 67–68.

24 Weiss 1978; Zubrod 1984; Braverman 1991.

BIBLIOGRAPHY

Baszanger, Isabelle, "Dechiffrer la douleur chonique: deux figures de la pratique médicale", *Sciences Sociales et Santé*, vol. 9 (1991), 31–78.

Baszanger, Isabelle, "From pain to person: a new object for medicine", pp 155–170 in I. Löwy (ed.), *Medicine and Change: Historical and Sociological Studies of Medical Innovation* (Paris & London: John Libbey, 1993).

Berg, Mark and Annemarie Mol (eds.), *Differences in Medicine*, (forthcoming).

Billings, John, "Address in medicine", *British Medical Journal*, vol. i, (1886).

Bosk, Charles, *All God's Mistakes* (Chicago: University of Chicago Press, 1992).

Braverman, Albert S., "Medical oncology in the 1990's", *Lancet* vol. i (1991), 901–902.

Collins, Harry M., "Researching spoonbending: concepts and practice of participatory fieldwork", pp. 54–69 in Collin Bell and Helen Roberts (eds.), *Social Researching: Politics, Problems and Practice* (London: Routledge and Kegan Paul, 1984).

Dodier, Nicolas, *L'expertise médicale: Essai de sociologie sur l'exercise du jugement* (Paris: Metaillé, 1993).

Emerson, Robert M., "Introduction", pp. 1–35 in R. M. Emerson (ed.), *Contemporary Field Research* (Illinois: Waveland Press, 1983).

Fintor, Lou, "Patient activism: cancer groups become vocal and politically active", *Journal of the National Cancer Institute*, vol. 83 (1991), 528–529.

Fleck, Ludwik, *Genesis and Development of a Scientific Fact* (Chicago: University of Chicago Press, 1979).

Geertz, Clifford, "Thick description: Towards an interpretative theory of culture", pp. 3–30 in Clifford Geertz (ed.), *The Interpretation of Cultures. Selected Essays* (New York: Basic Books, 1973).

Golinski, Jan, "The theory of practice and the practice of theory: sociological approaches in the history of science", *Isis* vol. 81 (1990), 492–505.

Latour, Bruno and Steve Woolgar, *Laboratory Life: The Social Construction of Scientific Facts* (London and Beverly Hills: SAGE, 1979).

Löwy, Ilana, "Variance of meaning in discovery accounts: the case of contemporary biology", *Historical Studies in Physical and Biological Sciences*, vol. 21 (1990), 87–112.

Löwy, Ilana, "Recent historiography of biomedical research" pp. 99–110 in Ghislaine Lawrence (ed.), *Technologies of Modern Medicine* (London: Science Museum, 1994).

Marks, Harry M., "Notes from the underground: the social organization of therapeutic research", pp. 297–335 in Russel C. Maulitz, and Diana E. Long (eds.), *Grand Rounds. One Hundred Years of Internal Medicine* (Philadelphia: University of Pennsylvania Press, 1987).

Moertel, Charles G., "On lymphokines, cytokines and breakthroughs", *Journal of the American Medical Association*, vol. 256 (1986), 31–41.

Negrier, S., T. Philip *et al.*, "Interleukin with or without LAK cells in metastatic renal cell carcinoma: a report of the European multicenter study", *European Journal of Clinical Oncology*, vol. 25 (suppl. 3) (1989), 21–28.

Panofsky, Erwin, *Renaissance and Renaissances in Western Art*, (New York: Doubleday, 1965).

Rosenberg, Steven A. *et al.*, "Observations on the systemic administration of autologous lymphokine-activated killer cells and recombinant interleukin-2 to patients with metastatic cancer", *New England Journal of Medicine*, vol. 313 (1985), 1485–1492.

Rosenberg, Steven A. *et al*, "Experience with the use of high doses interleukin-2 in the treatment of 652 cancer patients", *Annals of Surgery*, vol. 210 (1989), 474–485.

Snow, C.P., *The Two Cultures and A Second Look* (New York and Toronto: Mentor Books, 1963).

Traveek, Sharon, *Beamtimes and Lifetimes: The World of High Energy Physicists* (Cambridge, Mass.: Harvard University Press, 1988).

Weiss, David, "Animal models of cancer immunotherapy: Some considerations", pp. 101–109 in R. Lee Clark, Robert C. Hickey, and Evan M. Hersh (eds.), *Immunotherapy of Human Cancer* (New York: Raven Press, 1978).

Zubrod, C. Gordon, "Origins and development of chemotherapy at the National Cancer Institute", *Cancer Treatment Reports*, vol. 68 (1984), 9–19.

CHAPTER 7

The Living Scientist Syndrome: Memory and History of Molecular Regulation

Jean-Paul Gaudillière

> "But as to the facts of the occurrences of the war, I have thought it my duty to give them, not as ascertained from any chance informant nor as seemed to me probable, but only after investigating with greater possible accuracy each detail [...]. And the endeavour to ascertain these facts was a laborious task, because those who were eye-witnesses of the several events did not give the same reports about the same things, but reports varying according to their championship of one side or the other, or according to their recollection".
> (Thucydides, *History of the Peloponnesian War*, I-22).

INTRODUCTION

In 1988, a French textbook on "present time history" came off the printing press. The book documented an ever increasing number of articles dealing with post-World War II topics in prominent historical journals such as *Annales*, *La Revue Historique*, and *La Revue d'Histoire Moderne et Contemporaine* to argue that present-day history is not 'mere journalism' but a promising field for serious study.[1] Nonetheless, the authors listed specific methodological problems: the lack of time distance, the vast number of documents, the excessive influence of oral sources and personal accounts, the political and professional interests of historians who participate in the social worlds they investigate, and so on. For instance, the book points out that the number of pages of archival material stored in the Nixon's presidential library exceeds 40 million. In addition, the library keeps in reserve 4300 video tapes, 216 kilometers of films, 50,000 books and objects. During Nixon's six years in office, the documentary production of the White House amounted to roughly seven million pages per year.

Rather than symptoms of the unprecedented nature of contemporary science, rich documentary sources and numerous opportunities for historical studies seem to be simply characteristics of modern industrial societies. One may view many specificities often attributed to present-day science, i.e., the increasing number of professional scientists, the development of state research agencies, or the growth of large scale communication networks as characteristics of the social order which dominated Western countries during thirty years of economic growth and the Cold War. Since methodological issues faced by historians of contemporary science seemed to be shared by general historians, the responses and skills needed are not unique and specific debates are not necessary.

Although the historiography of contemporary science and technology has much in common with the history of industries, war operations, or information networks, I do think that a few specificities must be acknowledged. Many historians who conducted interviews or used written testimonies by scientists who are still alive have experienced the divide between memory and history. This widely discussed experience is especially baffling when historians interact with living scientists. To put it in a nutshell: scientists perceive their activities as the discovery of pure facts about nature. They tend to produce highly specific accounts of their work and to demand a special kind of historical treatment. They not only ask for commemorative and celebratory narratives like most human communities, but expect linear and logical tales about the unveiling of nature. This paper will reflect and discuss these tensions between scientists and historians by looking at our experience in the writing of the history of molecular biology in France and the United States.

TRACING THE INVENTION OF ALLOSTERIC PROTEINS

For two years, I collaborated with Angela Creager on a joint project dealing with the history of allosteric proteins, a topic in biological research which emerged in the 1960s and which attracted the attention of biochemists, physical chemists, and molecular biologists.[2] Molecular biology is far from being an uncharted historical territory. Historians have, however, seldom attempted to produce fine-grain analysis of laboratory practices.[3] One reason may be that, in contrast to nineteenth century evolutionary debates, contemporary issues in biological research have become too technical to be easily grasped. It may well be that a majority of historians in the field originally worked as scientists in molecular and cellular biology. Acculturation is a long process and we usually think that our new skills are better deployed when writing narrative on the cultural resources or the political economy of science. Having completed such a study of the making of molecular biology in post-war France, I became interested in detailed studies of laboratory work.[4] The decision to embark on a study of the invention of allosteric proteins originated both in local opportunities in the form of large collections of manuscripts and letters which were available at the Pasteur

Institute, as well as in the possibility of a comparative study focusing on similar research paths in the Virus Laboratory at the University of California, Berkeley.

Biology textbooks usually situate the discovery of allosteric proteins in the early 1960s at the Pasteur Institute as part of Jacques Monod's work on bacterial cell metabolism. A decade later, when the notion had been stabilized, allosteric proteins were viewed as key enzymes controlling physiological adaptation in living cells. Allosteric proteins then displayed features articulated in a highly theoretical model advanced by Monod and his collaborators in 1965. According to this scheme, allosteric proteins are large macromolecules composed of several subunits involved in the catalysis of one single biochemical reaction. These building blocks may combine in various ways so that the binding of small compounds can induce changes in the global structure and modify the catalytic properties of the enzyme. Consequently, physiological regulation can be achieved by activating or inactivating allosteric proteins. For instance, 'feedback' mechanisms which are operationally defined as the ability to stop the synthesis of a compound when it accumulates in the cell are currently believed to be mediated by allosteric enzymes.

By the early 1970s, the meaning of 'allostery' was therefore two-fold. On the one hand, it was a sign-post or a catch word used by many biologists. It merely meant 'metabolic regulation of protein function mediated by changes of structure'. Allostery was used as an overarching 'molecular' explanation of adaptive processes in living organisms. On the other hand, it was a rather esoteric specialty within the realm of protein studies focusing on the deciphering of changes in the structure of half a dozen proteins.

Allosteric proteins have proved an interesting topic for epistemological work which addresses the role of physics and chemistry in biological research and/or the reductionist views of molecular biologists.[5] In addition, the origins and fate of allostery indeed constitute an interesting challenge for sociologically-minded historians of science. Our own interests were three-fold. Firstly, Monod's innovation mobilized cultural resources which ranged from the local — but widely publicized — 'Pasteurian' culture of bacterial physiology to post-war cybernetics.[6] The nature of the links between allosteric regulation and the 'MIT-based cybernetic nexus', an important element in the post-war scientific mobilization, was an interesting issue. A second set of questions related to laboratory practices and the confirmation/refutation problem. The Pasteurian definition of allosteric entities changed three times in the 1960s while the list of enzymes to be reflected upon or used as experimental models varied. Since the list of allosteric enzymes determined the list of significant results that could be used to confirm (or oppose) the (Pasteurian) model, one may argue that the history of allostery is a story of taxonomic regress.[7] In other words, allosteric proteins were what the specialists called allosteric proteins.

But who was to be part of the 'core set'? Not surprisingly, the Pasteurian work and Monod's theoretical definition did not go unchallenged. Famous discussions on the mechanism of enzymatic regulation thus opposed the physical chemist

Daniel Koshland and the Pasteurian group. However, the most interesting aspect was that a few debates were actually opened. For instance, results from the Berkeley group, where people used different tools to study different enzymes were commonly used to confirm or complement the Pasteurian experiments and models. The third problem we wanted to address was the general acceptance of the new entities as key components of the culture of molecular biology in the late 1960s.

THE REWARDS OF EX-SCIENTISTS: MOBILIZING LABORATORY NOTEBOOKS

The choice of a comparative and collaborative study was a choice of standing. The decision was made to start a detailed comparison of the laboratory work that resulted in the making of allostery in Paris and Berkeley. Thus, we built on our scientific experience: one of us completed a PhD in Biochemistry which dealt with the properties of a famous allosteric enzyme, the other one was trained as a biochemical engineer. As former biochemists, our status was that of skilled observers interested in finding discarded documents such as merely drafts and laboratory notebooks.

Harold Garfinkel once defined ethnomethodological observation as an "attempt to detect some expectancies that lend to commonplace scenes their familiar, life-as-usual character, and to relate this to the stable social structures of everyday activities".[8] Social scientists trained in the ethnographic tradition have usually advocated for naive observers. In this context, 'naive' is not synonymous with 'unaware of technical issues' but suggestive of 'not participating in scientific work'. In other words, observers should not be involved in the management of technical material and the production of scientific meanings. Following this tradition, Michael Lynch wrote about his stay in a brain science laboratory:

> "During this fieldwork I attempted to learn as much as I possibly could [...] about the natural sciences practices that composed the lab's research and training programs. [...] Even within the narrow area I studied, my competence with electron microscopic work never approached a practitioner's skills [yet] limited competencies gave considerably more access to the talk and conduct which I witnessed in the lab than would have been possible had I relied solely on the analytical skills of a social scientist [...]".[9]

This methodological stance has resulted in some fascinating description of laboratories. Yet, it has proved of limited interest for addressing issues related to the stabilization of scientific knowledge. 'Laboratory studies' have indeed rapidly declined. Ethnographers who recently ventured into science settings have deliberately avoided the analysis of knowledge claims and controversies.[10] Insider's knowledge may be necessary for perceptive analysis.

I would like to illustrate this point by commenting on the use of research notebooks. Laboratory notebooks have been extensively employed by historians interested in scientific creativity.[11] As the immunologist, Peter Medawar put it "What scientists do has never been the subject of scientific [...] inquiry. It is no use looking to scientific papers for they not merely conceal, but actively distort the reasoning that goes into the work they describe".[12] Our goal was not to single out events later associated with discoveries, but to investigate contextual decision making. In other words, we were chiefly interested in laboratory notebooks and the insights they gave into the preliminary stages of the process of (re)constructing actions. We viewed notebooks as records which could document alternations, temporary choices, forgotten alternatives. The aim was to focus on time flow and analyze the paths towards irreversibility.

The problem may be illustrated with an example regarding the boundaries between fact and artifact. One important episode in the prehistory of allosteric proteins was the 'simultaneous discovery' of 'desensitization' in Paris and Berkeley. This was defined as the complete loss of sensitivity of an enzyme to regulatory molecules without significant loss of catalytic activity. This was exemplified with curves plotting the velocity of enzymatic reactions to the concentration of substrate: following treatments like heating or exposure to urea or mercury salts, the 'abnormal' S-shaped curve displayed by regulated enzymes were turned into 'normal' hyperbolic curves.

In the early 1960s, John Gerhart was a student at the Berkeley Virus Laboratory. He was working on a thesis dealing with the properties of an enzyme (aspartate transcarbamylase) viewed as a key element in the regulation of amino-acid synthesis in bacteria. Careful study of Gerhart's laboratory notebooks suggested that he had traced desensitization-like curves in March 1961. Treatments of the enzyme by heat or mercurials then became means to destroy 'feedback inhibition', i.e., to destroy sensitivity to physiological compounds without diminishing the catalytic activity of the enzyme. This led Gerhart to introduce the 'two sites theory' which formed the basis of later discourses on allosteric regulation. In a paper presumably written in July, Gerhart claimed that treatments with denaturating agents could separate regulatory effects from catalytic activity. This fact supported his argument that two different binding sites on the protein were responsible for these properties. The paper presented experiments with heat and mercurial compounds arranged in hypothetico-deductive reasoning:

> "*If* the inhibitor binding site is really specific and has little to do with binding the substrate, then disrupting this site should greatly reduce the inhibition, but should not necessarily affect the enzymatic activity. [...] Three *ways* have been found for destroying inhibition completely".[13]

Thus, one may claim that John Gerhart discovered 'desensitization' on 15 March 1961 and worked out the meaning of the new phenomenon over a few months. At the same time, Jean-Pierre Changeux, then a student of Monod, was doing inactivation experiments with another bacterial enzyme involved in amino acid synthesis.

Closer examination of Gerhart's notebooks suggested a different interpretation. Treatment by mercurials had been practiced at the Berkeley Virus Lab since September 1960 to support the opposite theory, namely that inhibitory compounds and substrates competed for the same site. For example, on 7 December 1960, Gerhart resumed bench work with assays of various mercurials on his enzyme. Measurements provided all kinds of relations: some mercurials inhibited the enzyme but did not modify regulatory inhibition; other mercurials were inhibitors when employed at high concentration but activators when employed at low concentration; others stimulated the enzyme but did not affect feedback inhibition. The organizing principle was to be found in chemical structures: "small mercurials [Hg and contaminants in CH_3Hg?] knock down [inhibition and] activity rapidly. Bigger ones knock out inhibition only. Biggest one [Klotz dye] knocks out neither".[14] These complex effects originated in procedures aimed at producing a chemical map which would show the chemical groups involved in catalytic activity and feedback inhibition. Accordingly, mercury compounds which would artificially abolish enzymatic function were handled like metabolites inducing physiologically relevant inhibition. Chemical artifacts and biochemical facts remained closely associated until the spring of 1961 when systematic kinetic measurements were introduced to complete chemical mapping with the effect of disentangling operational relations between denaturation and regulation.

Looking at Changeux's notebooks exemplified the contingency of Gerhart's path: at the Pasteur Institute denaturating agents were associated with purification procedures. They were artefacts thought to induce structural changes that had no bearing on physiological control. In addition, Changeux employed chemical mapping in the opposite manner, namely as a tool to distinguish the putative catalytic and regulatory sites. Thus, one could argue that there was no 'desensitization' but two sets of practices and results which resulted in different (but changing) boundaries between facts and artifacts. Preserved notebooks do not show the transformation of a contrasted sequence of work into one single interpretative framework. In this case, they simply suggest that homogenization originates neither in analogous practices, nor in direct collaboration, but in parallel literary choices made when Changeux and Gerhart wrote papers addressing the same biochemical audience.

When thinking about primary sources, most historians of science accept a classification scheme with laboratory notebooks at the top, textbooks and popularization articles at the bottom. This hierarchy suggests that the series — notebooks, letters, drafts of papers, articles, institutional reports, treatises — reflects a chronological order of (re)construction. David Gooding recently advocated that notebooks are "cognitive, real-time, and non-linear" reconstructions while research articles are "rhetorical, retrospective and linear" reconstructions that enable persuasion and dissemination.[15] His classification of various modes of scientific reconstruction may be used to apprehend the complex relationship between laboratory notebooks, other scientific texts, devices, and audiences:

Traces of scientific work and modes of reconstruction:

Mode of reconstruction	*Texts*	*Devices*	*Audiences*
Real time, nonlinear	Notebooks	Local constructs, embodied skills	Research team
Retrospective, non-linear	Letters, drafts	Prototypes, inscriptions, protocols	Informal networks
Retrospective, linear	Reports, articles	Standardized tools	Community of specialists
Prospective, linear	Articles, treatises, essays	Public displays	Students, lay audience

Insider knowledge is not simply a knowledge of tools employed by scientists, but is also a minimal understanding of the culture shared by the practitioners. For example, insights into the history of phrases such as 'feedback regulation' or 'servomechanisms' were quite valuable. The fact that signal theory, system analysis or process control have played an important role in shaping molecular biology and the study of biological regulation networks is widely acknowleged. However, scholars who have sought to trace transfers of engineering concepts to biological settings have had limited success. Norbert Wiener and the group briefly assembled under the patronage of the Macy Foundation have sometimes been credited with the linkage between information theory, cybernetics and biological problems.[16] In all probability, it did make a big impact on the biochemical scene. Although Wiener's *Cybernetics* was immediately translated into French, Monod's papers do not document any specific use of the book. Knowing that control systems and servomechanisms were developed by physicists and engineers in the context of military research and that MIT was an important link provided a clue.[17] This led us to investigate an informal network which circulated the culture of control and regulation, namely a group of biochemists interested in 'bacterial cell physiology' which included the inventor of 'feedback inhibition', Edwin Umbarger (Harvard Medical School), Boris Magasanik at MIT, Gerhart's mentor Arthur Pardee who moved from U.C. Berkeley to Harvard Medical School in 1962, the former physicists Aaron Novick and Leo Szilard, and Jacques Monod at the Pasteur Institute.[18]

THE REWARDS OF EX-SCIENTISTS: SECURING TRUST AND COOPERATION

Because of our mixed status as historian-biochemists, interviewees did not fear we would use their statements to write 'debunking sociology' or 'mud racking journalism'. The biologists we recognized as key players in the field agreed to participate in long interview sessions. Occasionally, our informants were keen on providing letters and laboratory notebooks. We were usually not perceived as 'historians', but rather as 'scientists' temporarily involved in the writing of a piece of 'our' history. Interviewees sometimes explained that they felt relieved not to be speaking to 'journalists' or 'philosophers'. Exactly what they meant is unclear. A good guess may be that it was not necessary for them to start painful explanations of the meaning of 'allosteric', for example.

The feeling that historians too are engaged in the establishment of 'facts' enhanced cooperation and trust. This was reinforced by the deployment of historian's skills, i.e., the ability to find and use unknown unpublished documents. One anecdote may be revealing. One informant once reminisced about a meeting with Monod and Jacob when he was a student looking for a research topic. During the interview, the interviewee told us that he turned down several research topics proposed by Monod. These putative projects focused on the regulation of gene activity in bacteria, then at the top of the pasteurian research agenda. Reluctantly, Monod mentioned new results in the study of enzymes. Our informant jumped at the idea and embarked on the study of feedback inhibition. At one point during the session, we produced a letter that Monod had sent to an American colleague three months before the reported meeting. The letter mentioned our informant, and his putative work on feedback inhibition. Monod asked for permission to work on the enzyme which our informant later 'decided' to study. The document was puzzling enough to make our interviewee worry about individual memory. Not surprisingly, the episode strengthened the notion that historians are researchers making 'discoveries' who may help scientists get the 'facts straight'.

These dynamics multiplied problems with ambiguous beliefs in 'objectivity'. Peter Novick has carefully analyzed the part played by the 'objectivity question' in the rise of the historical profession. As he recapitulated, early versions of the objectivity dream included "a commitment to the reality of the past, and to truth as corresponding to that reality; a sharp separation between knower and known, between fact and value, and, above all, between history and fiction".[19] Techniques involved in the making of historical objectivity, i.e., critical evaluation of conditions of production and cross-checking, commonly focused on the nature of written sources. Historians know that claims for a description of what 'really' happened are questionable. They know that there is nothing like one single past to be described. Historians select 'events' and 'meanings' in order to tell stories. Accordingly, they are willing to play with changing perspectives. Nonetheless, historians are not sociologists who may think that documents are of little importance, that interesting lines of reasoning and models constitute the core of the trade.[20] Documents have to be collected, established, criticized, and carefully interpreted. Current translations of the 'evidence' problem therefore consist of decisions regarding 'what should count as a meaningful piece of information', 'what is an acceptable ordering of sources' and 'who is going to read and use the story'.

WORKING WITH THE MOLECULAR BIOLOGISTS: THE INVENTION OF TRADITIONS

> "My play is a memory play, it's not realistic
> In memory, everything seems to happen to music
> This explains the fiddle in the wings"
> (Tennessee Williams, quoted in J.P. Changeux's laboratory notebook 1).

Once viewed as a means to gain access to events and people which did not leave historical records behind them, oral history is routinely employed by contemporary historians to complement written sources.[21] Critics have argued that this is the most dangerous usage of transcripts since one may unwisely juxtapose sources produced in contexts which may be decades apart.[22] Therefore, it would be historical *naiveté* to employ oral history as a source: it actually documents the making of collective memories. I would like to illustrate and qualify this statement in two different ways. Firstly, one may argue that written sources as well as oral sources are instrumental in 'inventing traditions'. Secondly, when compared with other material, one important specificity of oral history is the creation of a middle ground mediating between historians and participants.

Our best informant at the Pasteur Institute was Jean-Pierre Changeux who carried out numerous experiments on allosteric proteins in the early 1960s. In 1980, Changeux published a biographical account as part of a collective homage to Monod.[23] After our first meeting, he went back to his personal archives. He sent us a copy of his laboratory notebooks and spent a good deal of time analyzing their content. Then he wrote a new scientific biography and asked for comments. As became clear from the discussion, one motive for this interest in the primary sources was the attribution of priority in the discovery of allosteric proteins. His second account was published in a biochemical journal.[24]

Changeux's first account suggested that allostery was a theoretical discovery gradually made by Monod. The second text is a detailed description of experiments that led to the description of the basic features of allosteric regulation. This second narrative suggests that allosteric properties were discovered at the same time in Paris and Berkeley. The first account is an exercise in the memory of a founding father. The second is a serious, though sometimes selfserving, reconstruction of research paths. Insights into the making of scientific memory may be gained from a comparison of these texts. We felt at odds with both narratives since personal recollections often prevail over written sources. This practice resulted in tropes well known to students of experimental narratives.

Stories of bad management (for instance, storage of enzymes at the wrong temperature or unintended treatment by chemicals) are commonplace in our transcripts. Elements of specific sequences of work are presented as the product of pure serendipitous observation ("the properties of threonine deaminase *I used as the control* changed with time") which triggered new hypotheses and goals ("A reproducible method was necessary to abolish the regulatory properties of threonine deaminase without destroying the enzymatic activity. I tried replacing the aging of the enzyme by heat treatment".) This logic was impossible to cross-check. The aim of such a sequence, however, is not the analysis of specific courses of action but an example of the proper relationship between observation and hypothesis. These anecdotes provide vivid images of scientific open-mindedness.

Accounts of important conversations commemorate founding fathers or scientific legacies while they contribute to building up the culture of an emerging collective. The fate of de Gaulle's speech of June 18, 1940 is well-known by historians of contemporary France. In the midst of military debacle, a few people in France could listen to BBC programs. Nonetheless, an amazingly large number of would-be participants in the French Resistance remember the precise circumstances that led them to turn the radio on at the time when de Gaulle launched his appeal. When confronted with historian's skepticism, these actors usually stick to their memory. In like manner, Dominique Pestre, who worked on the history of CERN, once described high energy physicists' reactions to the findings of professional historians.[25] The claim that Niels Bohr was reluctant toward the establishment of the European laboratory was never accepted by CERN physicists even when they looked at original documents. Adjustments were made to ease the tensions between historical claims and collective memories of the founding father. Similar adjustments were made by our above mentioned informant confronted with Monod's letters. His second thoughts unambiguously restored the first account. Monod's letters could be used to argue that nothing like an open-ended discussion about our informant's research topic ever happened. The alternative was to change dates in order to have the putative meeting take place before Monod's writing. This was not wrong. It was just doubtful.

Observers of memory routinely emphasize the part played by displacement, generalization and coalescence in the making of life histories.[26] Similar processes contribute to the emergence of (local) scientific culture and may result in fairly different recollections of the same event. In 1980, Changeux started his account of Monod's invention of the 1965 model for allosteric proteins with a discussion of Changeux's experimental results. Then he "launched into a series of very general reflections about the structure of globular proteins. [...] He built models — out of cardboard at first, then with balls of clay, and then with dice — which helped him think like a geometer".[27] Theoretical insights led Monod to postulate that allosteric transitions preserved a putatively symmetrical arrangement of the protein components. This 'pure theorizing' was a highly praised component of the Pasteurian world: "The Anglo-Saxon audience, always very pragmatic, did not see the logical necessity, and also the experimental basis of the postulate of symmetry conservation". R.L. Baldwin, an American physical chemist, was then on sabbatical at the Pasteur Institute. He remembered Monod tinkering with cardboard models while Changeux worked out computer models of his kinetic studies. Rather than being a positive insight, the postulated symmetry was viewed as a constraint leading to an overly simple model. Monod's approach seemed strange to the physico-chemist:

> "I was intrigued by Jacques approach to the problem. First of all I was surprised that he expected to solve the problem of cooperative binding just by thinking about it and studying the clues in the literature. He did not consider it necessary to have training in physical chemistry".[28]

Oral history may be viewed as a special historical practice because historians shape interviewee's responses by the way their questions are phrased. A few changes introduced by Changeux in his second text about the prehistory of allosteric proteins exemplify this collective 'invention of traditions'. We were interested in documenting the continuity of work procedures in Monod's laboratory. Most questions about the local know-how triggered unfavorable comments. A few hints were however taken as good guesses. For instance, we suggested that Monod's early tinkering with chemicals viewed as analogs of the small metabolites inducing the synthesis of bacterial enzymes, may have informed Changeux's early experiments. The proposal reflected our interest in the 'reasoning with hands'. Our idea echoed Changeux's attempt to show that his study of allosteric proteins was a direct legacy of Monod's early work on enzymes and to neutralize the vivid memory of Monod's research in bacterial genetics. Changeux's laboratory notebooks did not reveal systematic use of substrate analogs. Nevertheless, this vision of the local continuity became a middle ground. Our analysis stressed a local line of reasoning. We emphasized the practice of buying, synthesizing, and screening analogs which emerged in the 1950s to be later reinvested in the quasi-pharmacological survey that Changeux organized in 1961.[29] In his second narrative, Changeux recollects Monod's mentorship. Following his entry into the lab, "he experienced, on the bench, the striking structural difference existing between substrates of the enzyme galactosidase and compounds that induce its biosynthesis". Our intuition was not only 'confirmed', but turned into a public 'fact' that could be quoted to support our original interpretation!

WORKING WITH THE MOLECULAR BIOLOGISTS: COMPETING DISCOVERY ACCOUNTS

As time went by, conflating aims turned the practice of oral history into an uneasy collaboration. Like most historians of science, we viewed laboratory notebooks as fundamental resources for analyzing past experiments. In contrast, our informants routinely relied on belated reconstructions when confronting notebooks, papers and memory. These contrasted hierarchies of sources originated in contending visions of scientific work.

Discovery accounts written by scientists are sometimes apologetic series of achievements arranged in some kind of chronological order. More complex narratives advocate for a vision of wrong paths, and failures. Though preconditions and recognition paths may be acknowledged, the vision of scientific work is usually naturalistic and psychological. The discovery process to be described remains retrospective, linear, highly theoretical, and takes place in an isolated space.[30] Changeux's second narrative aimed at explaining how the idea of allostery was first turned into a successful perspective at the Pasteur Institute. Descriptions were mainly informed by published articles which defined phenomena and settled priority debates. Two issues triggered competing uses of sources: the part played by techniques and the role of audiences.

This may be illustrated with a sensitive aspect of the rivalry between Paris and Berkeley, namely preliminary stages in the making of desensitization. Changeux's description is an ideal sequence of work. Thanks to serendipitous observation, Changeux showed "that 10 minutes at 55°C led to complete loss of sensitivity to isoleucine without significant loss of enzyme activity". Then, "I coined the term 'desensitization' to refer to a phenomenon which had also been observed in the USA by John Gerhart and Arthur Pardee". Discussion with Monod resulted in the creation of a model suggesting "the existence of two distinct sites which we would respectively designate as activity site and inhibition site". Finally,

> "one of the simplest predictions of this hypothesis was that there should be two groups of compounds, one analogous to isoleucine, and one to threonine [...] these analogs might, for instance, act as steric antagonists of either the substrate or the regulatory inhibitor".

Systematic survey of analogs was organized and the final classification confirmed the hypothesis.

In contrast, we stuck to the chronological order of the laboratory notebooks. So, results that could be viewed as early instances of desensitization when referred to the final paper, appeared as attempts to achieve a technical task, i.e., purify the enzyme. Thus, we paid much attention to instances of 'reasoning with hands'. Our story goes like this. The first thing Changeux achieved in order to domesticate his enzyme was to find out conditions for an activity test. Then, he followed two related aims: the 'technical' aim was to improve the purification process; the 'scientific' aim was to explain changes of activity and also 'disinhibition' which was then a technical problem to be circumvented. Disinhibition meant that Changeux was sometimes unable to obtain the physiological phenomenon, i.e., reduced activity following exposure to regulatory compounds. This was not a 'real' phenomenon, but an artifact apparently originating in the purification procedures. Disinhibition was, however, defined with the same kind of operations that made physiological inhibition meaningful. For some time, these patterns as well as other unstable effects of regulatory molecules, remained within the context of the purification problem. As time went by, however, the experimental system was stabilized in a different configuration. Treatments like thermal inactivation, or exposure to chemicals were routinized, and their meaning changed. Once used to follow up artefactual inactivation, they were employed to produce an enzymatic fraction no longer inhibited by isoleucine and a 'desensitization phenomenon'.

Technicalities, however, are not the whole story and literary achievements have a life of their own. Thus, feedback inhibition did not become allosteric regulation when Changeux worked out the 'two sites hypothesis' but later on, in the summer of 1961, as Monod wrote his "General Conclusions" of the Cold Spring Harbor Symposium on "Cell Regulation". This may be viewed as a natural theoretical move. An alternative approach is to focus on the interplay between the Pasteurian history and the disciplinary dynamics of the gathering. Yearly meetings at Cold Spring Harbor then served as significant socio-cultural events in the life of molecular biologists. What was gained by replacing the term 'feedback inhibition' with 'allosteric inhibition'? Was this simply a linguistic coup to link up all sorts of regulatory phenomena? Monod's first definition of 'allosteric' pointed out the lack of resemblance between substrates and inhibitors of enzymes. As mentioned above, it was a way to renew his biochemical work on substrates and inducers of enzyme synthesis which had been played down by virtue of his collaboration with Jacob and the rising study of genetic regulation at the Pasteur Institute. Allosteric properties were rooted in biochemical *habitus*. At the same time, they were useful in keeping a balance between biochemistry and genetics. Rather than settling marginal proprietary issues regarding the invention of 'feedback enzymes', allostery as a presumptive 'second secret of life' aimed at disciplinary benefits. As Jacob and Monod were the stars of the meeting with gene expression and bacterial genetics, Monod breathed new life into the biochemical audience with regulatory enzymes.

Discovery accounts, despite an overly rationalized and 'whiggish' tone, rarely opposed *local* archival material. Discrepancies and competing interpretations rather pertained to the relevant 'discovery context'. To our informants, allosteric enzymes respecting the 1965 model have always been out there. Consequently, whatever the local contingencies, someone would have discovered their properties in the 1960s: the discovery could have been made in Boston, if not in Paris or Berkeley. The relevant 'context' was therefore defined by the list of regulatory enzymes which were under investigation in Western biochemical laboratories. Our context was defined by the collective life of molecular biologists.

One interesting dimension of this contrast was the comparison of events in Paris and Berkeley. From the biologist's viewpoint, comparison was easy. John Gerhart and Jean-Pierre Changeux worked on different enzymes with different methods. Yet, they both worked on feedback regulation and proteins showing 'catalytic and regulatory binding sites'. Results were not replicated but confirmed by virtue of independent discoveries. From the historian's viewpoint, things were quite dissimilar. Firstly, we couldn't collect in either of the two settings documents dealing with simultaneous and identical issues. Secondly, our sources illustrated specific practices rather than common patterns. For instance, by virtue of the collaboration with the local physical chemist H. Schachman, the analytic ultracentrifuge became, in Berkeley, the most important tool for the study of allosteric proteins. In contrast, the Pasteurian team did not try to domesticate the machine. Thus, one may contrast Monod playing with cardboard models to Schachman's skills at modifying the ultracentrifuge, i.e., American engineering and Pasteurian theorizing.

Making the two laboratories comparable was a good historical practice, but it undermined naturalistic discovery accounts. As mentioned above, we built on the example of controversies studies. Gerhart's notebooks were used to trace alternatives to the pasteurian logic and *vice-versa*. This was instrumental in reviving local uncertainties. For instance, above, in 1963–1964, Monod redefined allosteric proteins according to the hemoglobin exemplar and initiated theoretical debates with physico-chemists working on the blood protein. This move led him to give up the 'two sites theory' as a basic feature, and to develop the 1965 kinetic model. Gerhart too became interested in similarities between allosteric enzymes and hemoglobin. The Berkeley team, however, employed hemoglobin metaphorically. Hemoglobin never became *the* allosteric protein: the blood oxygen carrier showed analogies with allosteric enzymes. In California, imported structural data on hemoglobin did not replace physico-chemical studies of enzymes. Beside variable shaping of local experimental systems, diversity of cultural and human resources presumably determined the disparity. Thus, we started to trace the different meanings of hemoglobin on both side of the Atlantic. As several informants argued, the methodological problem we faced was then to decide which 'context' should be taken into account. The definition of the boundaries of the inquiry seemed to be as contingent as the bench practices we described. One problem was, for instance, to decide what to do with science policies.

THE POLITICS OF SCIENCE AND THE POLITICS OF SOURCES

Since laboratory notebooks are seen to reflect elementary scientific events, reports and documents produced in peer-reviews committees, managerial offices, or governmental departments are considered as tardy and rhetorical texts. Our experience was sometimes to turn this classification upside down.

As allosteric enzymes were emerging at the Pasteur Institute, the French political system was in turmoil and science policy was in the midst of major redefinition. De Gaulle seized power in the spring 1958. One year later, the Fifth Republic and its powerful presidency were established. For years, prominent scientists including the Pasteurian biochemists had claimed that the country was unable to compete with the pace of research prevailing in the United States. Opposing the traditional elites and eager to push the modernization of the country, Gaullist officials were ready to consider the notion that most universities in the country were failing in the support of up-to-date scientific research.[31] The first government of the Fifth Republic thus agreed to a major boost in state funding for scientific research and industrial development. In addition, the prime minister established a new governmental agency, the Délégation Générale à la

Recherche Scientifique et Technique (DGRST), whose mission was to support applied research and technology.

Much to the surprise of the French biologists, the scientists and officials in charge of writing the list of DGRST goals picked up the development of molecular biology. Though state management of science was common practice in post-war France, DGRST was unusual. The committees established within this institutional framework were political bodies rather than peer-review bodies. Once a field was chosen as a target, a small group of scientists appointed by the Prime Minister would propose and implement action plans. They would make all decisions concerned either with the creation of new institutes, the establishment of fellowships or the management of research contracts. Once the decision to set up a special program for molecular biology was made, Monod and colleagues were put in charge.

In 1960 Monod wrote the draft of a general plan for the development of the discipline. His report focused on the relationship between proteins and nucleic acids. The text emphasized the recent progress in understanding gene action. Though the Pasteurians were strongly represented, the committee reflected a balanced choice of disciplines (microbiology, biochemistry, genetics, physical chemistry, immunology, and embryology), and of institutions (CNRS, universities, and private research institutions). One major issue was to define molecular biology and to decide who would benefit from funding and participate in the making of the new discipline. For complex reasons, the biochemists dominated the process. Thus, the study of enzymes and protein structure became a stronghold.

Five years later, a new plan was discussed that would define goals for a second program. Once again, Monod wrote most of the report. Emphasis then shifted from studies of genetic regulation to structural analysis of macromolecules. It is no surprise that the subunit structure of proteins and allosteric models were at the top of the agenda. As an exemplar of 'molecular cybernetics' allosteric regulation became the 'second secret of life' and it was presented as a putative clue to cell differentiation, hormonal action, developmental processes or nerve function. This perspective would be later expanded in conferences and in Monod's philosophical essay *Chance and Necessity*.[32] Before writing this report, Monod had been involved in the problematic writing of the 1965 model for allosteric proteins. The final version emphasized variable arrangements of subunits and 'allosteric transitions' while hemoglobin became the allosteric protein *par excellence*. The DGRST report echoed the paper.

There is nothing like one single causal link between the 1965 model, the politics of molecular biology, the fate of the democratic left in France, and the culture of control and regulation. Our first approach was to analyze the second DGRST report as a mere reflection of Monod's drafts. Later, policy documents were used to argue that hemoglobin was instrumental in linking the interests and practices of (medical) biochemists with the prospect of the pasteurian molecular biologists managing the DGRST program. This led us to change our hierarchy of sources and to take into account policy making choices when addressing Monod's theoretical choices.

Why did Monod jump at hemoglobin? By 1963, Monod was looking for something to renew the study of allostery. Changeux's experiments were stalling because of persistent purification problems. Attempts to establish alternative experimental systems proved of limited success: the Berkeley people refused to let the Pasteurians work with their enzyme and preliminary tests with other enzymes did not fulfill expectations. Monod was then deeply involved in science management and little time was left for experimental tinkering. To a superficial eye, hemoglobin seemed attractive because the physical chemists and the Cambridge molecular biologists had collected numerous data on the structure of the molecule. Hemoglobin was, however, many things to many people: investigations ranged from crystallography to pathology. When Monod adopted hemoglobin, he not only used the data from his Cambridge colleagues but he also imported a strategy. In Cambridge, hemoglobin had been turned into a 'mediating object' linking people doing X-ray crystallography and the biochemists deciphering amino acid sequencing.[33] In the French context, perspectives were broader and more political. By 1964, within the DGRST framework, hemoglobin already contributed to attracting biochemists into molecular biology. Firstly, it was a common topic for molecular biology seminars sponsored by DGRST as a means for (re)training. Secondly, important collaborative projects dealing with the structure and synthesis of the blood protein were supported by the agency.[34] In the same way, allosteric interactions, once associated to hemoglobin, could become 'boundary objects' linking heterogeneous scientific worlds, i.e. molecular biologists, biochemists, physical chemists, physiologists. Indeed one icon of the emerging (re)definition of allosteric proteins, the S-shape curves describing substrate binding, was routinely employed by physiologists and biochemists discussing oxygen transport. Allosteric hemoglobin could likewise be weakly structured in common use and strongly structured in local use. Consequently, our final contention was to emphasize this disciplinary and political relationship when interpreting the 1965 model. Not surprisingly, this was a source of disagreement.

WHO SPEAKS FOR HISTORY? WHO SPEAKS FOR SCIENCE?

> "Imagine a study of Christian religious work, conducted by a scholar in, say, the Middle Ages in Europe. The study outlines the process by which priests and nuns choose careers; details the fights for resources between various monasteries and abbeys; describes the impact of publishing technology on the dissemination of theology; explains how servants help keep the churches clean and orderly, and so on. [...] Although it is true that logically, such a study could proceed in tandem with, or even independent of, specific articles of faith, in a theocratic society a study of the mundane and menial aspects of religious work is heretical".
> (Susan Leigh Star, *Regions of the Mind*).[35]

In conclusion, I would like to echo Star's metaphorical description of the sociology of scientific work and suggest that the tensions historians experience when interacting with scientists in the writing of contemporary history of science are rooted in competing visions of science. When we wrote about allostery, we didn't write about an isolated and culturally neutral field. Though everything is not always connected to everything, our historical choices depend on moral, political, economical, and cultural commitments that contribute to defining what we are keen at studying. Our informant public accounts were tales about proteins which contributed to legitimating the present enterprises of molecular biologists. Our study of lives with and within allostery aimed at illuminating aspects of the work practices, social meaning and public understanding of molecular biology in the 1960s. In all likelihood, similar approaches of present day science as well as collaborations between historians and 'their' scientists-informants-readers-and competitors-in-the-writing-of-history would be much more difficult to achieve if the community of professional historians of science was to disappear. Yet, since our narratives are value-laden, should we wonder about who they are for? Professional historians and social scientists? Obviously. Science students? Yes. Wider audiences? That would be salutary, but how to avoid wishful thinking?

ACKNOWLEDGEMENTS

I am indebted to Ilana Löwy, Angela Creager, Thomas Söderqvist and the participants of the International Workshop on the Historiography of Contemporary Science, Technology, and Medicine, held in Göteborg, Sweden, September 1994, for comments and criticisms.

NOTES

1 Soulet *et al.* 1988.
2 This paper is an interpretation of the work done in collaboration with the French biologists. For the dynamics of the research, see Creager and Gaudillière, 1996.
3 One exception being H.J. Rheinberger's (1992) study of Zamenick's laboratory.
4 Gaudillière (forthcoming-1).
5 For an epistemological survey, see Debru 1984.
6 On the former, see Gaudilliere 1993 and Burian *et al.* 1988. On the latter, see Keller 1990.
7 Collins 1980.
8 Garfinkel 1967, 37.
9 Lynch 1985.

10 Traweek 1988. Indeed, the naive observer depicted in Latour (1987) could follow his scientist-informant up to the moment when inscription making devices were displayed. Then he silenced all complaints and questions.
11 Holmes 1985 and 1991; Gooding 1989.
12 Medawar 1969, 151.
13 Gerhart 1962.
14 Manuscript in the possession of the author.
15 Gooding 1991.
16 Kay 1994.
17 Heims 1980.
18 Bennett 1993. On MIT and postwar military-industrial research, see Galison *et al.* 1992.
19 Novick 1988, 2.
20 Pollack 1992.
21 Thompson 1988. See also de Chadarevian (this volume) and Soojung-Kim 1989.
22 Peschanski 1987.
23 Changeux 1980.
24 Changeux 1993.
25 Pestre, "L'histoire du CERN: entre usage politique et légitimation des pratiques du milieu" (unpublished manuscript).
26 Raphael 1980.
27 Changeux 1980, 199.
28 Baldwin 1979, 204.
29 Creager and Gaudillière (cf. above, note 2).
30 For analyses of the function of discovery accounts, see Löwy 1990 and Abir-Am 1985.
31 Picard 1990.
32 Monod 1970.
33 Chadarevian 1996.
34 Gaudillière 1996.
35 Star 1989.

BIBLIOGRAPHY

Abir-Am, P.G., "Themes, genres and orders of legitimation in the consolidation of new disciplines: constructing the historiography of molecular biology", *History of Science*, vol. 23 (1985), 73–98.

Baldwin, R.L., "Discussions about proteins", pp. 202–209 in André Lwoff and Agnes Ullman (eds.), *Origins of Molecular Biology: A Tribute to Jacques Monod* (New York: Academic Press, 1979).

Bennett, S., *A History of Control Engineering, 1930–1955* (London: Institution of Electrical Engineers, 1993).

Burian, R., J. Gayon, and D. Zallen, "The singular fate of genetics in French biology", *Journal of the History of Biology*, vol. 21 (1988), 357–402.

Chadarevian, Soraya de, "Sequences, conformation, information: biochemists and molecular biologists in the 1950s", *Journal of the History of Biology*, vol. 29 (1996), 385–416.

Changeux, J.P., "Une thèse avec Jacques Monod: préhistoire des protéines allostériques" pp. 197–208 in André Lwoff and Agnes Ullmann (eds.), *Les origines de la biologie moléculaire. Hommage à Jacques Monod* (Montréal: Etudes Vivantes, 1980).

Changeux, J.P., "Allosteric proteins: from regulatory enzymes to receptors", *BioEssays*, vol. 15 (1993), 625–634.

Collins, H., *Changing Order* (London: SAGE, 1980).

Creager, A. and J-P. Gaudillière, "Meanings in search of experiments and *vice-versa*: the invention of allosteric regulation in Paris and Berkeley", *Historical Studies in the Physical and Biological Sciences*, vol. 27 (1996), 1–71.

Debru, C., *L'esprit des protéines* (Paris: Hermann, 1984).

Galison, P. and B. Hevley (eds.), *Big Science: The Growth of Large Scale Research* (Stanford: Stanford University Press, 1992).

Garfinkel, H., *Studies in Ethnomethodology* (New York: Polity Press, 1967).

Gaudilliere, J-P., "Molecular biology in the French tradition? Redefining local traditions and disciplinary patterns", *Journal of the History of Biology*, vol. 29 (1993), 473–498.

Gaudillière, J-P., *Le colibacille, le biologiste et le plan: histoire de la biologie moléculaire en France* (Paris: Masson, (forthcoming-1).

Gaudillière, J-P., 'Messenger RNA and the biochemists: The making of a scientific network', *Journal of the History of Biology*, vol. 29 (1996), 417–445.

Gerhart, John C. and Arthur B. Pardee, "The enzymology of control by feedback inhibition", *Journal of Biological Chemistry*, vol. 237 (1962), 891–897.

Gooding, D., *Experiment and the Making of Meaning* (Dordrecht: Kluwer, 1989).

Heims, S.J., John von Neumann, and Norbert Wiener, *From Mathematics to the Technologies of Life and Death* (Cambridge: MIT Press, 1980).

Holmes, F.L., *Lavoisier and the Chemistry of Life* (Madison: University of Wisconsin Press, 1985).

Holmes, F.L., *Hans Krebs*, (New York: Oxford University Press, 1991).

Kay, L., "Wer schrieb das Buch des Lebens? Information und Transformation der Molekularbiologie" in M. Hagner, H.J. Rheinberger, and B. Wahrig-Schmidt (eds.), *Objekte, Differenzen und Konjunkturen* (Berlin: Akademie Verlag, 1994).

Keller, E.F., "Physics and the emergence of molecular biology: a history of cognitive and political synergy", *Journal of the History of Biology*, vol. 23 (1990), 389–409.

Latour B., *Science in Action* (Cambridge: Harvard University Press, 1987).

Löwy, I., "Variance of meanings in discovery accounts", *Historical Studies in the Physical and Biological Sciences*, vol. 21 (1990), 87–112.

Lynch, M., *Art and Artifact in Laboratory Science* (London: Routledge, 1985).

Medawar, P., *The Art of The Soluble* (London: Penguin Books, 1969).

Monod, J., *Le hasard et la nécessité* (Paris: Seuil, 1970).

Novick, P., *That Noble Dream. The "Objectivity Question" and the American Historical Profession* (Cambridge: Cambridge University Press, 1988).

Peschanski, D., "Questions à l'histoire orale", *Cahiers de l'IHTP*, vol. 4 (1987), 21–33.

Picard, J.F., *La République des Savants, la Recherche Française et le CNRS* (Paris: Flammarion, 1990).

Pollack, M., "L'entretien en sociologie", *Cahiers de l'IHTP*, vol. 21 (1992), 109–114.

Raphael, F., "Le travail de la mémoire et les limites de l'histoire orale", *Annales ESC*, vol. 35 (1980), 17–45.

Rheinberger, H-J., "Experiment, difference, and writing. I.: tracing protein synthesis", *Studies in the History and Philosophy of Science*, vol. 23 (1992), 305–331.

Rheinberger, H.J., "Experiment, difference, and writing. II: the laboratory production of transfer RNA", *Studies in the History and Philosophy of Science*, vol. 23 (1992), 389–422.

Soojung-Kim, A., "Oral history and the history of science: a review assay with speculations", *International Journal of Oral History*, vol. 10 (1989), 270–285.

Soulet, J.F. and S. Guinle-Lorinet, *Précis d'histoire immédiate: le monde depuis la fin des années 60* (Paris: Armand Colin, 1988).

Star, L., *Regions of the Mind* (Stanford: Stanford University Press, 1989).

Thompson, P., *The Voice of the Past, Oral History* (2nd ed., Oxford: Oxford University Press, 1988).

Traweek, S., *Beamtimes and Lifetimes: The World of High-Energy Physicists* (Harvard: Harvard University Press, 1988).

CHAPTER 8

Electric Memories and Progressive Forgetting

Skúli Sigurdsson

> "Memory threatens history when it becomes a distinct genre of remembering, when the experiences we have of the past become so multiform that we have different ways of remembering different pasts and no one can be said to enjoy primacy over the others. When history is only one way among many of recapturing the past it becomes questionable whether history is an experience of the past at all, whether history is not a mere fiction; a fiction because one way among many of remembering the same past always stands exposed to the accusation of being an invention".[1]

INTRODUCTION: THE PROBLEMS OF PROGRESS TALK

The 1960s saw the largest hydroelectric power project ever carried out in Iceland at that time: a hydropower station in the glacial river Thjórsá at Búrfell. It was originally 105 MW and was soon enlarged to 210 MW. In conjunction with the hydropower station a large aluminum smelter was built near Hafnarfjördur, a neighboring community of Reykjavík in the South West; it was owned by the Swiss company Alusuisse. The coming of Alusuisse to Iceland provoked a fierce political debate in the 1960s, a period when a coalition government of the center-right Independence Party and the Social Democratic Party was in power. Center-leftists and socialists criticized the form of ownership of the factory and the energy price which the Swiss firm paid. At the end of the 1960s, in an interview in *Tímarit Máls og menningar* (Journal of the Society for Language and Culture), the professor of geology and prominent cultural figure Sigurdur Thorarinsson said to the editor of the journal, Kristinn E. Andrésson:

> " 'Stop now my lord ...' We have so many energy sources, that I think it is justified to use some of them for energy intensive industries. Yet here we must proceed with great

> caution, so we will not lose control of these precious national resources. And I think that those assertions are barely credible, which are constantly being repeated, namely that we must sell the hydropower quickly to foreigners, because soon nuclear power will be so cheap, that hydroenergy will not be able to compete with it".[2]

Thorarinsson's words point to a dimension of the history of electrification of Iceland which is now forgotten, namely that the exuberant promise of abundant and cheap nuclear power in the 1950s and 1960s seemed to threaten the attractiveness of domestic hydroelectric power projects. This was seen as especially troubling in Iceland, where hydropower and energy intensive industries were considered an important element in the modernization of the country and diversification of the monoculture fishing economy.

There is something uncanny about Thorarinsson's words, for merely a quarter century later it seems outlandish that the question of nuclear energy mattered in debates about electric power and modernization in Iceland. These words could have been taken from the 1980s documentary film *Atomic Café*, but not from a leftist literary magazine in 1969. Little did I know, as a boy in the 1960s reading avidly about the adventures of Tom Swift, about the nuclear drill, the rocket or the race to the moon, that I was reading about the active concerns of my parents' generation.[3] In this chapter I want to explore the meaning of swift evanescent memories, some in my very own past, and what they imply for writing the history of contemporary technoscience and technological modernism. It has recently been argued for the study of the human genome project that it may be appropriate to describe it as fast science instead of big science.[4] I want to make this appellation mine and explore in the following pages the meaning of swift memories in the age of high technology.

The process of forgetting the past is aided by the way our language is constructed and how we domesticate human-made artefacts with cultural metaphors like progress. On the basis of my experience working on a commissioned history of the electrification of Iceland I would argue that progress talk in all of its forms is a major reason for the difficulty one encounters when one tries to unearth the recent technoscientific past; progress talk induces forgetfulness. Gillian Beer notes: "One of the most remarkable powers of the human mind — less often commented on than its power to proliferate senses — is its power to exclude, or suppress, feasible meanings".[5] Progress with its deterministic connotations enshrines the present at the cost of the past and naturalizes the omnipresent technological environment in which we live. Here one might extend a concept which Ian Hacking has advanced and think of the styles of reasoning that accompany technological systems and contribute to their self-authenticating character.[6] Styles of reasoning are also styles of forgetting.

I begin the chapter by discussing the history of the electrification of Iceland and three aspects of memories that it unravels — atomic, radio and corporate — before I ponder, in the penultimate section, whose memories in general the historian should bring to light and in particular whose memories the historian of the

Icelandic power system should disclose. I conclude by discussing progress, technological thought styles and memory.

ATOMIC MEMORIES[7]

Condorcet's *Esquisse d'un tableau historique des progrès de l'esprit humain* is a classic statement of the idea of progress, yet it is more a sociology of error than a sociology of progress.[8] In other words, most of the the devotees of progress know from what they are fleeing rather than where they are heading. In the United States this attitude found expression in the belief in the frontier and an endless technological advance. Langdon Winner observes: "Rooted in colorful dreams, myths, and fantasies, expressed in recurring rituals that contain an enormous spiritual energy, the nation's involvement with technology is, among other things, clearly a matter of religious zeal".[9]

Insatiable need for energy characteristic for the post-war era proved a fertile ground for the propaganda for nuclear energy in the 1950s and 1960s. This was the period of rapid economic growth, when the gross national product (GNP) in the United States rose by 4 to 5 percent annually instead of 2 or 3 percent as in the post-1970 period.[10] The nuclear power advocacy responded to shifts in Cold War policies that came in the wake of the Soviet A-bomb in 1949, namely nuclear weapons were now perceived to be less useful for settling global military conflicts. Instead attention focussed on the symbolic display of nuclear technology as a sign of military and ideological superiority. In December 1953 President Eisenhower announced in a speech to the United Nations the program of "Atoms for Peace" and the revised Atomic Energy Act of 1954 encouraged the Atomic Energy Commission to encourage development of nuclear power by private companies and utilities. By stressing domestic consumption and abundant sources of energy the homely qualities of nuclear energy were emphasized and its association with national security was lessened.[11] This is not to say that domestic settings could not be a geopolitical arena, as the famous 1959 kitchen debate between Richard Nixon and Nikita Khrushchev demonstrates; the two men stood in front of a washing machine and argued about the virtues of the respective political systems.

The ethos of US nuclear advocacy found strong expression in the early 1960s in a book on energy in the Time-Life natural science series. In a chapter about the magic world of Albert Einstein the wonders of nuclear energy were presented, and although problems with the storage of radioactive waste were mentioned, the presentation was eerily optimistic. In a picture of the dumping of nuclear waste in concrete containers off the US coast, the photo caption assured the reader that the disposal site had been chosen after thorough oceanographic research and that the containers were supposed to last for a hundred years. The book appeared in Icelandic translation in 1968, five years after its initial appearance.[12]

This was a late outburst of nuclear euphoria in Iceland; in the 1950s the fledgling physics community and the electric power sector had found common interest in the atom and studies of nuclear energy. This was done with the establishment of a professorship in physics at the University of Iceland in the mid-1950s and the founding of a committee on nuclear physics which was financially supported by the electric power sector.[13] The Icelandic system builders wanted to be kept abreast of the latest developments in power technologies lest their precious hydropower resources were rendered useless, and for a while they expected that geothermal energy could be utilized to produce heavy water for export. Moreover, nuclear technology seemed to offer the hope for a backward country to leapfrog into the technological future without having to traverse all the intermediate steps from an agricultural past like its neighbors. Finally, in the buoyant 1950s the system builders thought that small nuclear power reactors might be employed to solve local energy problems in Iceland. Thus they sought bids for a small nuclear reactor in order to solve the energy supply problems of Vestmannaeyjar, an island fishing station south of Iceland (and site of a volcanic eruption in 1973).[14]

In the wake of the launching of Sputnik in 1957 a colorful maverick engineer, Gísli Halldórsson, wrote a book about space travel whose chief protagonists were Wernher von Braun, Albert Einstein, and rockets. The book was an ode to freedom:

> "With travel in rockets, in which solar or nuclear energy were used to propel the rocket, with electrons or photons, it might perhaps be possible to do away with time altogether. To displace oneself on the spacetime manifold. To glide out of the wheel tracks of time and find a short cut to the future!"[15]

These words are emblematic for the optimism of the post-war era in spite of the recent ravages of world history. In Iceland this progressive optimism fell on particularly fertile soil. Not only had the lighting power of electricity struck a responsive cord, but once the process of secularization, which had started in the late nineteenth century was completed, progress became *ersatz*-religion. It has been observed, that "It is truer of the votaries of progress than of the adherents of any of the great religions that they believe without knowing either quite what they believe or why they believe in it".[16] If hard pressed the average Icelander might have answered like the engineer Halldórsson: the goal was to reach wealth and bliss freed from the fetters of material constraint — but through the agency of energetic materialism.

Where had the technological momentum been directed after World War II? David Joravsky asks: "Did the benign technology of electrification engender the great malignancy of the warfare state?" He hopes the answer is no, yet he finds disturbingly compelling the evidence that Thomas P. Hughes presents in *American Genesis* for the "irresistible product of 'technological momentum' ".[17] One example of irresistibility is the closely connected development of nuclear submarines and nuclear power reactors that Hughes discusses at length in his book.[18] In this example the transfer of a military technology into a 'peaceful' domain worked well, whereas the fate of nuclear powered ships demonstrates that this process did not always work. Nuclear submarines are to the period after World War II what diesel powered submarines were to the period around World War I.[19] The employment of diesel engines in submarines boosted their development, as was made evident in the 1920s in the fishing village of Akranes north west across the bay Faxaflói from Reykjavík, where a small diesel motor was installed for producing electricity. It came from a French submarine, in which it had been used to pump air.[20]

Traditionally the Icelanders have seen hydropower as one of their main natural resources and sources of electric power, yet this is misleading. It was the availability of mass-produced diesel generators that made much of the electrification of Icelandic coastal towns possible in the 1910s and 1920s. In 1930, when the total production of electric power was 4,519 h.p. mainly used for lighting (3.4 MW; national population 109,000), about three-quarters was hydroelectric and the rest was generated with diesel engines. If the hydroelectric production for the two largest municipalities, Akureyri in the North and Reykjavík in the South West, is subtracted from this figure, the proportion of diesel generation rises significantly. These numbers are taken from a report by Jakob Gíslason, who became the director of the State Electric Control Office in 1933.

It was only after World War II that the Icelandic state entered the electric power sector actively; between the wars and during World War II it had done some preliminary engineering assessments but had primarily enforced safety regulations. In a companion report Gíslason had demonstrated that the number of fires caused by electricity had risen exponentially in previous decades, like the production of electric power itself. Between the wars the largest player in the electric power sector was the Reykjavík Electric Utility, which had built a small hydropower plant in the river Ellidaár, some seven kilometers away from the town center, at the beginning of the 1920s and enlarged it in 1933 (1921: 1 MW; 1923: 1.7 MW and 1933: 3.2 MW; population of Reykjavík was 18,000 in 1920 and 28,000 in 1930). Four years later a much larger hydroelectric power plant in the river Sog 44 kilometers from Reykjavík, the outlet of the lake Thingvallavatn, went into operation (1937: 8.8 MW; population of Reykjavík in 1940 was 38,000). This meant that suddenly the electricity supply in Reykjavík increased fourfold. In order to sell the surplus energy the Utility encouraged the installation of electric cooking stoves, but until then the majority of households in Reykjavík cooked with gas supplied by the Reykjavík Gas Works, which had been established in 1910. Although the Gas Works supplied most of the energy for cooking needs in Reykjavík until 1937 and continued to be in operation until the mid-1950s, this is a fact that seems to have been almost erased from the memory of those that grew up with it. This has been my experience speaking with people who were born in the 1920s and grew up in Reykjavík; paths no longer traveled become paths that were never traveled.

RADIO MEMORIES

In 1943 the Association of Electric Utilities was established. It signified the growing importance of electrification in Iceland and showed that the need had arisen for the various municipal electric utilities scattered along the coast to join forces. The structure of the Association was modeled after similar organizations in Scandinavia. At the beginning of the 1940s there were 40 electric utilities in Iceland, 25 of which satisfied the membership criteria of the new Association; the remainder were too small to merit inclusion. The primary tasks of the Association were, for example: the establishment of uniform procedures for collecting statistical data, establishment of technical norms, solving difficulties of importing electric power equipment during wartime, testing and calibration of electricity meters, and reducing radio disturbances.

The State Electric Control Office had distributed radio disturbance complaint forms nationwide and sent inspectors to municipalities with good results. The Reykjavík Electric Utility employed an inspector responsible for checking and recording radio disturbances that could result from electric wires in houses, electric gadgets (for example, irons, electric drills and fluorescent lamps), the radios themselves, antennas or inlets from the street. Radio disturbances are typical of the novel problems that come in the wake of the expansion of technological systems. Thus already in 1915, when the electric utility in the fishing station Vestmannaeyjar started operation, the State Telephone and Telegraph Board was concerned about disturbances from the new electric wires.

Radio disturbances continued to be discussed at annual Association meetings into the 1950s. The situation improved as the strength of radio broadcasting was increased and the number of relay stations grew. The State Radio did not commit resources to dealing with the problem and claimed it was the task of the electric power sector to solve it. At the fourth annual meeting of the Association in 1946 an extensive report was presented by the Reykjavík Electric Utility on radio disturbances: 41 % originated in the radios themselves and 59 % from the electric power network and home appliances. Thus it was argued that the State Radio should bear some of the cost of solving this problem. The report highlights the institutional resources required for gathering and accurately recording data about technical phenomena such as radio disturbances. In retrospect, it also illuminates Icelandic and international history.

One source of radio disturbances during the war was the growing number of electric drills, which reflected the increase of construction activity and prosperity that came in the wake of the occupation of Iceland by British and US armed forces during World War II. To account for the impact of the tremendous influx of foreign capital and the massive changes it wrought in the social fabric remains to be done in Icelandic economic history — the citizenry has conveniently forgotten the foreign salvation and the economic plight of the 1930s.[21] In the words of a contemporary observer and a future member of the engineering elite the changes in Icelandic society were momentous:

> "The prime mover in all these changes has, of course, been the tremendous capital flow of the war years, which has come like manna from heaven into this capital deprived country. Now for the first time in history there is sufficient domestic capital to start several enterprises, which in normal times would only have been conceivable had foreign venture capitalists thought it likely that they could make some money from it. Now the plan is to use some of these extensive funds to renew and build from scratch the fishing industry, on which the whole financial state of the country depends".[22]

It is difficult to think back to the time of World War II, when the war effort nearly drowned in the noise of its own technology, as Paul N. Edwards has argued. The noise of rockets, airborne bombs, and various combat vehicles was added to the sounds of World War I, namely those of artillery and gunfire. The high noise levels called for altered communication channels inside the combat vehicles so that the mixed human-machine life forms, the cyborgs, could remain operative.[23] The drowning noise of the battlefields had its counterpart in the importance of radio communication in other realms of the war effort and daily life. Paul Fussell observes:

> "In a way not easy to imagine in the present world of visual journalism, the war was mediated and authenticated by spoken language, whose conduit was the radio. For those at home the sound of the war was the sound of the radio. Actually, wartime was a special moment in the history of human sensibility, for in those pre-television days the imagination was obliged to fill in the missing visual dimension, and in those pre-tape days, there was in addition all the excitement of live transmission, when anything could happen".[24]

The importance attached to fighting radio disturbances in the early years of the Association of Electric Utilities is a striking confirmation of Fussell's remarks. The annual reports of the Association included transcriptions of discussions during meetings, technical presentations were reprinted and soon they included accounts of early electrification efforts, which are invaluable for the historian of technology as an example of an ongoing project of creating a collective memory. Thus the reports are not merely a unique source for capturing the sensibilities of the period, moreover they convey a vivid impression of the life of the Association and how its members forged a common identity that bound them together.[25]

The Association met in 1953 in the vicinity of lake Thingvallavatn, that is in the national park Thingvellir, the site of the old parliament. The historical recollections presented dealt with the history of electrification in Hafnarfjördur, Keflavík and Thingvellir. The first electric generator in Keflavík in the South West was introduced in 1922 by the fishing entrepreneur Matthías Thórdarson and was primarily used for illumination of his small fishing plant during the winter months. From 1923 until 1933 a small company supplied the inhabitants of Keflavík with electricity, whereafter the township took over operations. The number of users grew slowly and electricity was mainly used for lighting. The small electric diesel generators employed could meet the demand because the users took care to turn off the lights as they left each room. After 1940 the consumption of electricity grew dramatically and soon difficulties were caused by insufficient generating power. The war had started, people had more money and were no longer inclined to save electricity. This problematic situation lasted until Christmas 1945, when electric power from the Sog plant which had been enlarged in 1944 was delivered to Keflavík and the surrounding communities. The high voltage transmission line was built by the State Electric Control Office, the predecessor organization of the State Electric Power Works established in 1947.[26] In the post-war era the airport near Keflavík constructed during the war became the focal point of a large US base that brought economic benefits in a variety of ways: construction work, employment on the base, and lucrative electricity sales after 1960. Moreover, in the 1960s it was the US armed forces television and radio at the base that was pivotal in introducing rock'n'roll and other forms of foreign mass culture to Iceland.

CORPORATE MEMORIES

The examples in the preceding sections illustrate the various purposes memory can serve in large technological systems and for the formation of the affective forces that bind its members together. This brings to mind the variety of means employed historically to develop ingenious mnemonic techniques, be it in the sixteenth-century library of Michel de Montaigne, in the eighteenth-century pedagogical children's literature of Johann Siegmund Stoy or in the twentieth-century haunting picture boxes and collages of Joseph Cornell.[27] Jakob Gíslason's assessment of electrification at the beginning of the 1930s offered a detailed tabulation of the various local arrangements nationwide, and a kind of epidemiological analysis of the rise of fires caused by electrical leaks, the search for causes of radio disturbances in the early 1940s was similarly of an epidemiological nature. This is a statistical view of society, which indicates the pervasiveness of what Ian Hacking calls a statistical style of reasoning.[28] Resources had to be employed in order to gather these data and physical inscription techniques had to be used in order to preserve the statistical marks.

In the present, mnemonic techniques are of an electrical nature in the form of magnetic tapes and disks for storing characters that are threatened with obsolescence in the future, not only because of the fragility of the physical media and the danger of outdated technologies, but because of problems of interpreting the stream of characters correctly.[29] The load curves so essential for running large electric power networks can now be stored and recalled in far greater detail than was possible previously when the cost of storing the countless requisite data points was prohibitive.[30] This has interesting implications for the social history of technologial systems. Now it becomes conceivable to analyze the metamorphosis of the load curve landscape in minute detail; for example, the shift of the lunchtime peak into the evening hours for electric power consumption, so revealing of the changes in family life in recent years. Furthermore, the load curve landscape makes it possible to analyze the behavior of the population during a certain time frame. This is a statistical geography of society, lost is detailed information about the grain of sand, pebbles and rocks that make up society.

The reliable access to data in a large technological system is of a crucial value. In order to bill customers data must be gathered and stored, in order to project future growth past consumption trends must be analyzed and valleys in the load landscape identified. It would not be an exaggeration to say that a large electric power network is both harnessing data and energy resources, for in the absence of fast and secure data management the enterprise gradually grinds to a halt. Furthermore, large technological systems use information control to maintain their autonomy and market power and to resist institutional change. In a recent review of Carl Pechman's *Regulating Power: The Economics of Electricity in the Information Age* this point has been stressed:

> "Computer models are used in regulation and governance of the industry, as well as in operation and planning of the electrical system. Information control is one of the main sources of market power used by the utilities against the new competition in electricity generation. The traditional utilities have used computer models to improve their control of information and gain market power in both the regulation and operation of the industry".[31]

Although the modern corporation may be magnetically wired, it must also rely on traditional archival techniques for keeping track of letters, memos, records of meetings, etc. The National Power Company, founded in 1965 as a joint venture of the Icelandic State and the City of Reykjavík, built the hydropower station at Búrfell in the river Thjórsá. Looking back on the construction experience in the early 1970s, the civil engineer Jóhann Már Maríusson, currently deputy director of the National Power Company, noted that the paper flow generated had exceeded what hitherto had been known in construction projects in Iceland.

> "A rough calculation shows, that the supervision engineers wrote about 5,000 letters, the contractor Fosskraft wrote a bit more, between six and seven thousand letters, and other letters that ended in the custody of the supervision engineers from several sources connected with the project, were of the order of 5,000".

Furthermore, when it was taken into account that each letter was issued in five to eight copies and that besides the letter exchange several hundred reports were written, it was clear that much time had been consumed by writing and reading. Once the construction was finished a long litigation process began between the National Power Company and the contractor, Fosskraft. Then the reservoir of documents proved to be indispensable, although, as Maríusson wryly noted, the main difficulty proved to be how to interpret the documents and make their contents comprehensible to outsiders of the construction process.[32]

Maríusson's account is not only highly interesting for the light it throws on the complexities in the construction and running of large technological systems, and the strong interests attached to accurate memories of the past. It also shows that the information explosion is not solely a problem for scientists.[33] Maríusson began his account by giving an historical overview of the sequence of events that led to the establishment of the National Power Company in 1965 and the Búrfell project. The historical overview canonized the history of the company's genesis and emphasized that electric power in Iceland was synonymous with hydropower. The completion of the Búrfell project and the completion of the historical mythology coincided. This can be seen when Maríusson's nearly identical account of the history of electrification in Iceland two decades later is compared; myth has become history.[34]

This view of the history of electrification is widely shared as can be inferred from a recent article in a tourist magazine, *Iceland Review*, which discusses the electric hydropower projects in the Thjórsá region and a related environmental controversy in the late 1960s. The controversy resulted from plans to build a large reservoir in the oasis Thjórsárver near the southern edge of the icecap Hofsjökull that would have destroyed the nesting place of the pink-footed geese that come there every spring.[35] The initial reaction of the National Energy Authority was to claim that the geese had to be moved, for otherwise the economic feasibility of hydropower developments in the Thjórsá region would be seriously undermined. The Authority stressed this for example, in answering the International Wildfowl Research Bureau.[36] Yet, the reservoir was not built and eventually the geese were saved. This was the first moment of environmental awakening for Icelandic engineers, the second was the controversy about further extending the hydropower project in the river Laxá in the North, an outlet from the lake Mývatn, a renowned paradise for ornithologists. In an article in 1970 about the Laxá crisis, the chief engineer of the Authority, formerly its director Jakob Björnsson, defended the technocratic view of the controlled domination of nature and indicated that soon the development of nuclear power might make hydroelectricity obsolete; therefore the Icelanders had to weigh their options carefully.[37] A quarter century later, the memory of the public protest and sabotage that prevented the completion of the Laxá project is still vivid, whereas the notion of a possible nuclear challenge is beyond recall.

WHOSE MEMORIES?

Aldous Huxley agreed with the Cambridge philosopher C.D. Broad that the function of the brain, nervous system and sense organs was in the main eliminative and not productive. The function of the brain and nervous system was to protect us from being overwhelmed and confused by a mass of largely useless and irrelevant knowledge, by shutting out most of what we should otherwise perceive or remember at any moment, and leaving only that very small and special selection which was likely to be of practical use.[38] Friedrich Nietzsche had reached a similar conclusion in *History in the Service and Disservice of Life*, where he spoke about the animal that lived unhistorically and merged into the present like a number without a remainder. It knew nothing of dissembling, hid nothing and seemed always to be what it was; it could not help being honest. The predicament of man he described as follows:

> "But he wonders about himself, too, wonders that he cannot learn to forget, but always clings to the past; that however far or fast he runs, that chain runs with him. It is a wonder that a moment now here, now gone, a nothing before, a nothing afterward, still returns like a specter to disturb the calm of a later moment. Loose sheets from the scroll of time are forever falling out, fluttering away — and suddenly floating back into man's lap. Then the man says, 'I remember', and envies the animal who immediately forgets and sees each moment really die, lapse back into fog and night and disappear forever".[39]

If it is unhealthy to remember too much, what position should historians take? Should they as a professional group base their legitimacy on remembering or should forgetting, which may be dangerously close to self-annihilation, be an integral part of their repertoire of self-authentication? Do historians really have a free choice, for after all acts of forgetting and remembrance are central to the exercise of power.

Here it may be helpful to think of the function of remembering in contemporary society. 1994 was a good year for the memory industry in the West: 50th anniversary of D-day, 50th anniversary of the failed assassination attempt on Adolf Hitler, and 25th anniversary of the Woodstock rock festival. 1995 turned out to be good as well: 50th anniversary of the end of World War II, twentieth anniversary of the fall of Saigon, and 100th anniversary of the discovery of X-rays.[40] One sometimes wonders what history would look like in a different numerical system, say in base eight or if we chose to define the beginning of a century differently? (It began in 1900 in Germany and 1901 in France.[41] So mark your calendar for the beginning of the next millennium.)

Some of these anniversaries have caused uproar and much discussion. The history of World War II continues to haunt the Germans, whereas the same is less true of the Japanese.[42] The anniversary of the Hitler assassination attempt has shown how difficult it is to create a common memory for east and west Germans, to openly acknowledge the role of socialists and communists in the resistance to the Nazi regime and the role of the wives of the men who took part in the attempt.[43] The dropping of the atomic bombs on Hiroshima and Nagasaki has recently caused a tremendous debate in the US.[44] When one thinks of ravaged countries in recent memory it becomes clear how delicate and important it is to settle the scores with the past in order to move into the future. Here one is reminded of Northern Ireland, South Africa, the torture ridden countries of Latin America, and the Middle East.[45] In the case of Israel the remembrance of the Holocaust is a sensitive issue: it needs to be remembered, yet it is imperative that the society be neither defined through the eyes of the perpetrators nor the memory of the Holocaust be exploited for political ends.[46] Avishai Margalit writes concerning the need to remember the Holocaust:

> "The question of remembrance, however, remains important. We must remember not only the death chambers but also the rich complexity of Jewish life in Europe which died in them. Zionists resist reviving the memory of the life they rebelled against, but remembering the Holocaust requires a reevaluation of Zionist criticisms of the Jewish world that was destroyed. As for the form of remembrance, the only appropriate way to perpetuate the memory of the Holocaust is through meticulous documentation and mastery of facts. In the case of the Holocaust, the Devil is in the details".[47]

Denials of the Holocaust make clear how important it is to keep this memory alive.[48] Thus there are events that should never be forgotten, which constitute a limit to the idea that forgetting can be a realistic option for historians.

Yet the question of forgetting will not go away, for the mass of historical facts accumulated must be sorted out and most of it eliminated by some principle before it can be analyzed. The examples in the previous sections demonstrate the abilities of institutions to register and remember. The challenge for the historian is to bring the load curves and other quantitative inscriptions to life rather than to be overwhelmed by the sheer size of these enterprises, and to resist the seductive ideology of progress and betterment of life that is writ large in enterprises of this kind. Yet, the freedom that the commissioned historian of institutions has to define autonomously her or his task depends on the contractual relationship with the sponsors.[49]

In my experience of writing a commissioned history of the electrification of Iceland, what surprised me most was the wide gulf between the sponsors' expectations and my own, between those of the executives who seemed at times to be ordering yet another piece of electric power equipment and those of the historian of technoscience who resisted such co-optation and approached the issues completely differently. Neither of us understood the nature of the contract which we had initially signed: incommensurability comes in many kinds and not only in epistemology. They wanted the key facts but had not envisioned an interpretative social history; they wanted to celebrate the 50th anniversary of the establishment of the Association of Electric Utilities and became very dissatisfied when I failed to meet the contractual deadline; they wanted celebration but had not carefully contemplated the deep contradiction between the relativism of historical inquiry and cherished beliefs like progress.[50]

Gradually I began to understand that the electric power executives and I inhabited different worlds. They were of an earlier generation, which had pinned its hopes on progress in the post-war era, on industrialization and modernization. I belonged to a generation that had reaped the benefits of the spectacular post-war economic growth but had begun to question the premises of that growth and thought of the history of electrification in terms of ecological damage and nuclear disasters, or in more sanguine moments in terms of popular culture and rock'n'roll. In my frequent moments of doubt I found solace in Eric Hobsbawm's words regarding the demise of the *ersatz*-religion of progress:

> "Once upon a time, say from the middle of the nineteenth century to the middle of the twentieth, the movements of the left — whether they called themselves socialist, com-

> munist, or syndicalist — like everybody else who believed in progress, knew just where they wanted to go and just what, with the help of history, strategy, and effort, they ought or needed to do to get there. Now they no longer do. In this respect they do not, of course, stand alone. Capitalists are just as much at a loss as socialists to understand their future, and just as puzzled by the failure of their theorists and prophets".[51]

I felt compelled to write a story that questioned some of the basic premises of modernism and addressed post-modern times. To do that and preserve my intellectual autonomy in the service of the powers that be turned out to be an exhaustive and frustrating, yet eventually worthwhile experience; in the end I was given free rein and a new and very generous contract. Yet ultimate power resides with the electric power executives for besides my book they have decided to publish a traditional account of the history of the electrification of Iceland.[52]

The protracted tug of war between me and the executives was about power and memories. They were the custodians of the memory of the meaning of electrification in Iceland codified in the annual reports of the Association of Electric Utilities. For some thirty years the Association had tried to publish a history of electrification to no avail, which meant that the publication effort in the early 1990s was accompanied by considerable urgency while the historical recollections had become increasingly ceremonial. One of the reasons for the publication impasse was that the question of how to achieve closure had not been settled nor could it be in light of the colossal changes that had occurred in the power sector since the late 1960s with the rise of the National Power Company to a dominant position at the cost of the Reykjavík Electric Utility and the State Electric Power Works. Closure was impossible at the corporate level.

Could any single player be given primacy in the story, or should one write a story where the ideology of progress and corporate legitimacy was conspicuously absent? Should one emphasize the swiftness of the societal changes that had accompanied electrification and the unexpected outcome of the whole enterprise? Should one stress that a discourse about conserving and reproducing the past is also a discourse about what it means to be normal in the present, as Michael S. Roth has argued? Should one highlight that "public concerns with memory and with the investigations of it can be considered screens on which a culture projects its anxieties about repetition, change, representation, authenticity, and identity".[53] What does it mean that in the closing decades of the twentieth century interest in the workings of memory has become widespread, say in studies of multiple personalities and cases of repressed memories of traumas of sexual abuse?[54] Ian Hacking wonders skeptically "why has it been essential to organize so many of our present projects in terms of memory?", he wonders analytically "what are the dominating principles that lock us into memory as an approach to so many of the problems of life, from child rearing to patriotism, from aging to anxiety", and he wonders "what constructions underlie these principles?"[55] Could it be that we speak so much of memory because there is so little of it left, as Pierre Nora has argued?[56]

What did I remember from my past and for what purpose, and what could I remember as a professional historian? I grappled with these questions while I argued with the electric power executives and moved further and further away from their concerns, while simultaneously gaining an understanding of the inner workings of their life world that constituted an integral part of my youth and childhood. The story told in the previous three sections indicates what my answer turned out to be.

CONCLUSION

One way of understanding the notion that progress talk is an important mechanism for constantly and swiftly re-writing the past from the vantage point of the present may be gained, if one compares Hacking style large scientific revolutions with small scale Kuhnian revolutions. After a Hacking revolution there is a different feel to the world, it has acquired a new texture.[57] What is true for a large-scale Hacking revolution holds even more for massive technological changes; the time before and after the changes is partly incommensurable. One need only think back to the nineteenth century before the advent of electricity to see this point, or to compare the times before and after the coming of the automobile. In this sense Hacking revolutions correspond to progress in the large sense, whereas small Kuhnian revolutions correspond to incremental changes that may add up to massive changes.

Progress is an important rhetoric strategy for enforcing the domestication of such massive changes and telescoping the past into the present, novel thought styles that accompany such Hacking revolutions and technological changes are another. These styles are an essential factor in the re-textualizing of each time period; they determine how we think about right and wrong, and their self-authenticating character makes it all too easy to forget about pre-style ways of thinking.[58] The challenge to the historian is to break loose from the reigning thought styles of the present and to try to envision the passage from the past to the present and across revolutionary divides in all their bewildering abruptness and complexity. In the case of large technological systems this is not merely an academic exercise; it has undertones of survival value and may provide an answer as to why one should care to remember the technoscientific past at all. Winner writes:

> "In the late twentieth century a great many people — scholars and ordinary citizens alike — have begun to realize that the key question is not how technology is constructed, but how to come to terms with ways in which our technology-centred world might be reconstructed. Faced with a variety of social and environmental ills, there is growing recognition that what is needed is a process of redirecting our technological systems and projects in ways inspired by democratic and ecological principles".[59]

Progress is a deterministic notion and the challenge for the historian of technoscience is to highlight the realistic possibilities for technological choice so that humans may gain freedom and control over their technological destiny. The fate of the former Soviet Union is a tragic reminder of the calamity that results from a rigid and dogmatic approach to technology.[60]

In order to steer clear into the technoscientific future the historian needs to remember the past imaginatively, stress what are the options for control and convey this to the public in terms which it understands. Here the entangled meaning of memory may be appropriate, for it may enable the historian to speak directly to the populace at large. Everyone remembers and has memories.[61] When writing about the history of electrification, the progress technology par excellence, it may be particularly useful to approach the past in such manner and emphasize the unnaturalness of the ubiquitous light switch on the wall. The narrative needs to be critical and underscore that one of the enduring myths of the present, the myth of progress, is both a poor platform for human action and the writing of history.

ACKNOWLEDGMENTS

I thank Michael Becker, David Cantor, Soraya de Chadarevian, Michael Hagner, Kai Handel, Lisa Herschbach, Agatha Hughes, Hotze Mulder, Sybilla Nikolow, Monika Renneberg, Hans-Jörg Rheinberger, Sven Th. Sigurdsson, and Thorsteinn Vilhjálmsson for valuable criticism and help, and Thomas Söderqvist for his patience and advice.

NOTES

1 Motzkin 1994, 115.
2 Andrésson 1969, 360.
3 Appleton 1959, 1964 and 1966.
4 Fortun (forthcoming); also Daston (forthcoming) and Winner 1989.
5 Beer 1987, 41–42.
6 Hacking 1982, 1985, 1992a and 1992b; also Staudenmaier [1985] 1989, 192–201.
7 For an introduction to the history of electrification in Iceland see Björnsson 1952, Krabbe 1946, 222–245, and Sigurdsson 1995.
8 Baker 1975, 358.
9 Winner 1989, 49; also Staudenmaier 1989 and Lasch 1989.
10 Lemann 1994; also Joerges 1994.
11 Smith 1991; also Hughes 1989, 421–442.
12 Wilson *et al.* [1963] 1968, 163.
13 Magnússon 1987.
14 Kristinsson 1959; also Kristinsson 1958.
15 Halldórsson 1958, 169.
16 Ginsberg 1973, 636.

17 Joravsky 1989, 14.
18 Hughes 1989, 421–442; also Hughes 1987.
19 Renneberg 1993 and Funk 1916.
20 Sigmundsson 1954, 128.
21 Nordal 1994.
22 Sigurdur Jóhannsson to Pétur Thorsteinsson, 3 December 1945 (letter in author's possession).
23 Edwards 1994, Ch. 9.
24 Fussell 1989, 180–181.
25 Analysis of the Association's annual reports might be modeled on the analysis of the journal *Technology and Culture* by John M. Staudenmaier [1985] (1989).
26 Gudjónsson 1953.
27 Ophir 1991, te Heesen 1995 and McShine [1980] 1990.
28 Hacking 1982 and 1985.
29 Rothenberg 1995.
30 Hughes 1983.
31 Davies 1993, 8; also Hvelplund 1995.
32 Maríusson 1973.
33 Söderqvist (this volume).
34 Maríusson 1994; also Mali 1994.
35 Kristjánsson 1993.
36 *Orkumál 19* (June 1969), 95–96.
37 Björnsson 1970.
38 Huxley [1954, 1956] 1977, 19.
39 Nietzsche [1874] 1990, 88.
40 Kragh 1995, 1–3.
41 Bennett *et al.* 1994, 1.
42 Craig 1994 and Buruma 1995; also Kramer 1995 and Ash (forthcoming).
43 E.g., Dempsey 1994, and Meding 1994.
44 E.g., Harwit 1994.
45 E.g., Krauthammer 1994.
46 Elon 1993.
47 Margalit 1994, 10; also Bartov 1995.
48 Lipstadt 1994, 30–33.
49 Cantor 1992, Krige and Pestre 1987, and Pestre 1990.
50 Abir-Am 1992, 323.
51 Hobsbawm [1978] 1984, 282.
52 Not all historians are so fortunate, cf. Cantor 1992, 132 and Hackmann 1993, 171–172.
53 Roth 1989, 49.
54 Hacking 1994 and Schacter 1995.
55 Hacking 1995, 3.
56 Nora 1989, 7.

57 Hacking 1987 and Fortun and Schweber 1994, 328–329.
58 Hacking 1992a and 1992b.
59 Winner 1992, 450; also Winner 1986 and 1989.
60 Graham 1993 and Holloway 1993.
61 Thomas 1995.

BIBLIOGRAPHY

Abir-Am, Pnina, "A historical ethnography of a scientific anniversary in molecular biology: the first protein x-ray photograph (1984, 1934)", *Social Epistemology*, vol. 6 (1992), 323–354.

Andrésson, Kristinn E., "Um jardvísindi og fleira. Rabbad vid Sigurd Thorarinsson [About geology and more: Conversation with Sigurdur Thorarinsson]", *Tímarit Máls og menningar*, vol. 30 (1969), 348–364.

Appleton, Victor, *Ævintýri Tom Swifts. Kjarnorkuborinn* [The Adventures of Tom Swift: The Nuclear Drill] (Skúli Jensson trans., Hafnarfirdi: Bókaútgáfan Snæfell, 1959).

Appleton, Victor, *Ævintýri Tom Swifts. Eldflaugin* [The Adventures of Tom Swift: The Rocket] (Skúli Jensson trans., Hafnarfirdi: Bókaútgáfan Snæfell, 1964).

Appleton, Victor, *Ævintýri Tom Swifts. Kappflugid til tunglsins* [The Adventures of Tom Swift: The Race to the Moon] (Skúli Jensson trans., Hafnarfirdi: Bókaútgáfan Snæfell, 1966).

Ash, Mitchell G., "Becoming normal, modern and German (again?)", in Michael Geyer (ed.), *Public Culture in Contemporary Germany* (forthcoming).

Baker, Keith M., *Condorcet: From Natural Philosophy to Social Mathematics* (Chicago: University of Chicago Press, 1975).

Bartov, Omer, "An idiot's tale: memories and histories of the Holocaust", *The Journal of Modern History*, vol. 67 (1995), 55–82.

Beer, Gillian, "Problems of description in the language of discovery", pp. 35–58 in George Levine (ed.), *One Culture: Essays in Science and Literature* (Madison, WI: University of Wisconsin Press, 1987).

Bennet, Jim, Robert Brain, Simon Schaffer, Heinz Otto Sibum and Richard Staley, *1900 The New Age: A Guide to the Exhibition* (Cambridge: Whipple Museum of the History of Science, 1994).

Björnsson, Glúmur, "Islands kraftförsörjning", *Tímarit Verkfrædingafélags Íslands*, vol. 37 (1952), 25–48.

Björnsson, Jakob, "Um náttúruvernd med sérstöku tilliti til virkjana [On environmental protection with special attention to hydroelectric power projects]", *Samvinnan* vol. 64:2 (1970), 22–27.

Buruma, Ian, "The unforgiven", *Time* (September 11, 1995), 66–67.

Cantor, David, "Contracting cancer: the politics of commissioned histories", *Social History of Medicine*, vol. 5 (1992), 131–142.

Craig, Gordon A., "An inability to mourn", *The New York Review of Books* (July 14, 1994), 43–45.

Daston, Lorraine, "The vertigo of scientific progress", in Bruce Mazlish (ed.), *Looking Back to the Future: Fins-de-siècles, 1900 and 2000* (forthcoming).

Davies, Andrew, "Idealizing the market: neglecting alternatives in the supply of energy", *EASST Newsletter* 12:4 (1993), 5–9.

Dempsey, Judy, "Germany split over legacy of war resistance", *Financial Times* (July 20, 1994), 2.

Edwards, Paul N., *The Closed World: Computers and the Politics of Discourse in Cold War America* (Cambridge, MA: MIT Press, 1996).

Elon, Amos, "The politics of memory", *The New York Review of Books* (October 7, 1993), 3–5.

Fortun, Michael, "Projecting speed genomics", in Michael Fortun and Everett Mendelsohn (eds.), *The Practices of Human Genetics: International and Interdisciplinary Perspectives* (Sociology of the Sciences Yearbook, vol. 19) (forthcoming).

Fortun, Michael and Silvan S. Schweber, "Scientists and the state: the legacy of World War II", pp. 327–354 in Kostas Gavroglu, Jean Christianidis, and Efthymios Nicolaïdis (eds.), *Trends in the Historiography of Science* (Boston Studies in the Philosophy of Science, vol. 151) (Dordrecht: Kluwer, 1994).

Funk, Gustav, "Die Dieselmaschine", *Tímarit Verkfrædingafélags Íslands*, vol. 1 (1916), 41–46.

Fussell, Paul, *Wartime: Understanding and Behavior in the Second World War* (New York: Oxford University Press, 1989).

Ginsberg, Morris, "Progress in the modern era", pp. 633–650 in Philip P. Weiner (ed.), *Dictionary of the History of Ideas, vol. 3* (New York: Charles Schribner and Sons, 1973).

Graham, Loren, *The Ghost of the Executed Engineer: Technology and the Fall of the Soviet Union* (Cambridge, MA: Harvard University Press, 1993).

Gudjónsson, Valtýr, "Saga rafveitu Keflavíkur" [History of the Keflavík Electrical Utility], *Ársskýrslur Sambands íslenzkra rafveitna*, vol. 11 (1953), 173–180.

Hacking, Ian, "Language, truth and reason", pp. 48–66 in Martin Hollis and Steven Lukes (eds.), *Rationality and Relativism* (Cambridge, MA: MIT Press, 1982).

Hacking, Ian, "Styles of scientific reasoning", pp. 145–165 in John Rajchman and Cornel West (eds.), *Post-Analytic Philosophy* (New York: Columbia University Press, 1985).

Hacking, Ian, "Was there a probabilistic revolution 1800–1930", pp. 45–55 in Lorenz Krüger, Lorraine J. Daston, and Michael Heidelberger (eds.), *The Probabilistic Revolution I: Ideas in History* (Cambridge, MA: MIT Press, 1987).

Hacking, Ian, "The self-vindication of the laboratory sciences", pp. 29–64 in Andrew Pickering (ed.), *Science as Practice and Culture* (Chicago: University of Chicago Press, 1992a).

Hacking, Ian, " 'Style' for historians and philosophers", *Studies in the History and Philosophy of Science*, vol. 23 (1992b), 1–20.

Hacking, Ian, "Two souls in one body", pp. 433–462 in James Chandler, Arnold I. Davidson, and Harry Harootunian (eds.), *Questions of Evidence: Proof, Practice, and Persuasion Across the Disciplines* (Chicago: University of Chicago Press, 1994).

Hacking, Ian, *Rewriting the Soul: Multiple Personality and the Sciences of Memory* (Princeton: Princeton University Press, 1995).

Hackmann, Willem, [Review of Joan Lisa Bromberg, *The Laser in America, 1950–1970*, 1991], *Isis*, vol. 84 (1993), 171–172.

Halldórsson, Gísli. *Til framandi hnatta* [To Distant Planets] (Reykjavík: Almenna bókafélagid, 1958).

Harwit, Martin, "The Hiroshima debate stills divides Americans, 49 years on", *The International Herald Tribune* (August 8, 1994), 6.

Hobsbawm, Eric, "Should poor people organize?" pp. 282–296 in Hobsbawm, *Worlds of Labour: Further Studies in the History of Labour* (London: Weidenfeld and Nicolson, [1978] 1984).

Holloway, David, "The politics of catastrophe", *The New York Review of Books* (June 10, 1993), 36–38.

Hughes, Thomas P., *Networks of Power: Electrification in Western Society, 1880–1930* (Baltimore: Johns Hopkins University Press, 1983).

Hughes, Thomas P., "The evolution of large technological systems", pp. 51–82 in Wiebe E. Bijker, Thomas P. Hughes, and Trevor J. Pinch (eds.), *The Social Construction of Technological Systems: New Directions in the Sociology and History of Technology* (Cambridge, MA: MIT Press, 1987).

Hughes, Thomas P., *American Genesis: A Century of Invention and Technological Enthusiasm 1870–1970* (New York: Penguin, 1989).

Huxley, Aldous, *The Doors of Perception & Heaven and Hell* (London: Granada, [1954, 1956] 1977).

Hvelplund, Frede, "(Dis)ability to change: energy conservation in the Danish electricity industry", pp. 205–221 in Arne Kaijser and Marika Hedin (eds.), *Nordic Energy Systems: Historical Perspectives and Current Issues* (Canton, MA: Science History Publications, 1995).

Joerges, Bernward, "Ein Zeitalter der Energie", pp. 106–133 in Wolfgang Zapf and Meinolf Dierkes (eds.), *Institutionenvergleich und Institutionendynamik* (Berlin: WZB-Jahrbuch/Edition Sigma, 1994).

Joravsky, David, "Machine dreams", *The New York Review of Books* (December 7, 1989), 11–15.

Krabbe, Thorvald, *Island og dets tekniske Udvikling gennem Tiderne* (Copenhagen: Gyldendal, 1946).

Kragh, Helge, "Scientific anniversaries", *International History of Science Newsletter* (April 1995), 1–3.

Kramer, Jane, "Letter from Germany: the politics of memory", *The New Yorker* (August 14, 1995), 48–54, 56–63, 65.

Krauthammer, Charles, "Reconciliation: Mandela finds a sound model in Chile", *The International Herald Tribune* (September 13, 1994), 6.

Krige, John and Dominique Pestre, "The how and the why of the birth of CERN", pp. 523–544 in Armin Hermann, John Krige, Ulrike Mersits, and Dominique Pestre (eds.), *History of CERN*, vol. 1 (Amsterdam: North-Holland, 1987).

Kristinsson, Björn, "Skýrsla um verd raforku frá kjarnorkustödvum [Report on electricity prices from nuclear power stations]", *Ársskýrslur Sambands íslenzkra rafveitna*, vol. 16 (1958), 228–231.

Kristinsson, Björn, "Orkuverd frá litlum kjarnorkustödvum" [Energy prices from small nuclear power stations], *Tímarit Verkfrædingafélags Íslands*, vol. 44 (1959), 53–57.

Kristjánsson, Jón Gudni, "Liquid power", *Iceland Review*, vol. 31 (1993), 34–40.

Lasch, Christopher, "The idea of progress in our time", pp. 229–239 in Steven L. Goldman (ed.), *Science, Technology, and Social Progress* (Research in Technology Studies, vol. 2) (Bethlehem: Lehigh University Press, 1989).

Lemann, Nicholas, "Mysteries of the middle class", *The New York Review of Books* (February 3, 1994), 9–13.

Lipstadt, Deborah E., "Denying the Holocaust: the fragility of memory", *Brandeis Review* (spring 1994), 30–33.

McShine, Kynaston, *Joseph Cornell* (New York: The Museum of Modern Art, [1980] 1990).

Magnússon, Magnús, "Kjarnfrædanefnd Íslands" [The Icelandic Committee on Nuclear Power], pp. 77–127 in Thorsteinn I. Sigfússon (ed.), *Í hlutarins edli: Afmælisrit til heidurs Thorbirni Sigurgeirssyni prófessor* (Reykjavík: Menningarsjódur, 1987).

Mali, Joseph, "Narrative, myth, and history", *Science in Context*, vol. 7 (1994), 121–142.

Margalit, Avishai, "The uses of the Holocaust", *The New York Review of Books* (February 17, 1994), 7–10.

Maríusson, Jóhann Már, "Virkjun Thjórsár vid Búrfell. 1. áfangi 3 x 35 MW" [Construction of the power plant Búrfell in the river Thjórsá: 1st phase 3 x 35 MW], *Tímarit Verkfrædingafélags Íslands*, vol. 58 (1973), 66–73.

Maríusson, Jóhann Már, "Raforka og virkjun vatnsfalla á Íslandi" [Electricity and the harnessing of hydropower in Iceland], *arkitektúr verktækni skipulag* vol. 15:3 (1994), 47–51.

Meding, Dorothee von, "Unscheinbar, privat", *tageszeitung* (July 20, 1994), 3.

Motzkin, Gabriel, "Memoirs, memory, and historical experience", *Science in Context*, vol. 7 (1994), 103–119.

Nietzsche, Friedrich, *Unzeitgemässe Betrachtungen/Unmodern Observations*, William Arrowsmith, ed. (New Haven: Yale University Press, 1990).

Nora, Pierre, "Between memory and history: *les lieux de mémoire*", *Representations*, vol. 26 (1989), 7–25.

Nordal, Jóhannes, "Vöggugjafir lýdveldisins" [The republic's gifts at birth], *Fjármálatídindi*, vol. 41 (1994), 145–155.

Ophir, Adi, "A place of knowledge re-created: the library of Michel de Montaigne", *Science in Context*, vol. 4 (1991), 163–189.

Pestre, Dominique, "Some characteristic features of CERN in the 1950s and 1960s", pp. 783–807 in Armin Hermann, John Krige, Ulrike Mersits, and Dominique Pestre (eds.), *History of CERN*, vol. 2 (Amsterdam: North-Holland, 1990).

Renneberg, Monika, "Wissenschaft, Wirtschaft und Politik — Gründung und Aufbau des GKSS-Forschungszentrums in Geesthacht", NTM: *Naturwissenschaft, Technik, Medizin*, new series, vol. 3 (1993), 147–159.

Roth, Michael S., "Remembering forgetting: *maladies de la mémoire* in nineteenth-century France", *Representations*, vol. 26 (1989),49–68.

Rothenberg, Jeff, "Ensuring the longevity of digital documents", *Scientific American* (January 1995), 24–29.

Schacter, Daniel L., "Memory wars", *Scientific American* (April 1995), 107–111.

Sigmundsson, Jón, "Upphaf og thróun rafmagns á Akranesi" [The beginning and development of electricity in Akranes], *Ársskýrslur Sambands islenzkra rafveitna*, vol. 12 (1954), 125–138.

Sigurdsson, Skúli, "Iceland electrified", *Uppsala Newsletter* no. 23 (fall 1995), 3–5.

Smith, Michael, "Advertising the atom", pp. 233–262 in Michael J. Lacey (ed.), *Government and Environmental Politics: Essays on Historical Developments Since World War Two* (Baltimore: Johns Hopkins University Press, 1991).

Staudenmaier, John M., *Technology's Storytellers: Reweaving the Human Fabric* (Cambridge, MA: MIT Press, [1985] 1989).

Staudenmaier, John M., "Perils of progress talk: some historical considerations", pp. 268–293 in Steven L. Goldman (ed.), *Science, Technology, and Social Progress* (Research in Technology Studies, vol. 2) (Bethlehem: Lehigh University Press, 1989).

te Heesen, Anke, *Das Werkzeug des Begreifens. Die Bilder-Akademie für die Jugend (1780–84) des Johann Siegmund Stoy*, Doctoral dissertation, University of Oldenburg, 1995.

Thomas, Keith, "Retrochic", *London Review of Books* (April 20, 1995), 7–8.

Wilson, Mitchell, *et al.*, *Orkan* [Energy], Páll Theodórsson transl. (Reykjavík: Almenna bókafélagid, [1963] 1968).

Winner, Langdon, *The Whale and the Reactor: A Search for Limits in an Age of High Technology* (Chicago: University of Chicago Press, 1986).

Winner, Langdon, "Technological frontiers and human integrity", pp. 48–64 in Steven L. Goldman (ed.), *Science, Technology, and Social Progress* (Research in Technology Studies, vol. 2) (Bethlehem: Lehigh University Press, 1989).

Winner, Langdon, "Social constructivism: opening the black box and finding it empty", *Science as Culture*, vol. 3 (1992), 427–452.

CHAPTER 9

Knowledge of the Brain: The Visualizing Tools of Contemporary Historiography

Susan E. Cozzens

The modern phenomenon of converging research areas presents one of the challenges of doing contemporary history. Contemporary biomedicine, which consumes nearly half the world's scientific and technical research talent, is a vast network of interrelated topics, with a structure that shifts constantly in response to instrumental and conceptual advances. Many of the major areas of advance that are ripe for historical analysis are therefore also quite large, with no distinct boundaries and highly complicated internal structures. The historian must either take an artificial slice out of this complexity, small enough to understand but too small to be of much significance, or attempt to grasp patterns of change in wide expanses of scientific knowledge.

In my work on the emergence of the interdisciplinary neuroscience movement in the United States, I have faced this challenge. In this chapter, I sketch where the problem of understanding the research agenda of a large field stands in relation to the overall issue of understanding its evolution, and what tools we have available for addressing it.

The version of this problem I face, as a present-day student of science, technology, and society (STS), is much more complicated than the problem faced by, for example, an old-fashioned intellectual historian. As an STSer, I approach the study of history with the assumption that science, politics, and social order have all been *co-produced*[1] through processes of negotiation and the creation of meaning in specific social interactions. I have no choice about whether to include in my concept of the neurosciences the web of human effort, inside and outside laboratories, that produces 'neuroscience knowledge'. Given that commitment on my part, the easy research strategy would be to focus on something small and local, where I can watch the interactions. For example, I could follow a scientist through her daily life,[2] and display the interactions within which culture, politics, and social relations are woven into her research results. I could then follow those results out of her laboratory, as a new set of interactions incorporates them into the daily worlds of many other people.

Unfortunately for me, I have actually chosen quite a different strategy in my project on the neurosciences. My research incorporates the assumption that neuroscientists construct laboratories and facts out of the cultural and political resources at their disposal, thus providing knowledge of a particular sort to be used in the construction of larger social and political patterns by other actors. Nonetheless, I have chosen to develop an overview of the rapid growth of this vast field rather than a close-up of some particular laboratory or research group. My students and I have been studying the institutional anchoring of the neuroscience movement, through the histories of professional associations, funding programs, and university departments of neuroscience. We have been tracing the incorporation of neuroscience knowledge into popular culture and into the repertoire of neuro-active drugs available and used in various American communities. Linking these two, we have attempted to grasp the emerging problem structure of the neurosciences, something we call 'the research agenda'. We treat all of these phenomena as the changing elements of co-production, of a modern technoscience.[3]

The larger question I wish to raise in this chapter is: Why bother with the overview? Is the language that one uses to describe large swaths of the network of interaction so constraining that one loses sight of the interactions themselves in the process? Conversely, does one *see* anything distinctive at the aggregate level, or is it merely the sum of smaller patterns of interaction? As you will see from my discussion, the effort involved in creating the overview is considerable. Is it worth it?

My chapter begins with the institutional structure that brain researchers created through their interactions starting in the 1960s, to give you a more concrete sense of my research object. I then turn to the theoretical problem that inspired my work on the neurosciences, the relationship between autonomy and power in a contemporary scientific movement. In particular, I make a case for the importance of grasping the research agenda as a key link in understanding the other elements of co-production. Finally, I review the tools available for visualizing the research agenda, along with my own frustrations in using them. In the end, I return to the question: Is the pain of creating an overview worth the results?

THE INSTITUTIONAL GROWTH OF THE NEUROSCIENCES

What we now call neuroscience — i.e., the interdisciplinary investigation of brain and behavior — has existed for at least a century.[4] But the laboratory activity has taken different institutional forms at different times over the last one hundred years. The story of the neurosciences that I am trying to tell starts at the end of what I have called elsewhere the Age of Institutes, and the beginning of the Age of Associations.[5] In the early twentieth century, in both the United States and Europe, brain research institutes were established as safe havens for fundamental research, a new organizational form where researchers could interact without the constraints of disciplinary teaching departments. Nonetheless, most U.S. brain researchers remained either anatomists or physiologists, not neuroanatomists or neurophysiologists.

The event that marked the beginning of a new institutional era for brain research was the formation in 1959 of the International Brain Research Organization. The idea for this organization was formulated at an international physiological conference in Moscow. Herbert Jasper, a Canadian brain researcher, took up the project and, with significant help from others, negotiated its initial funding from UNESCO. The organization grew as an elected body of researchers from many countries. It began organizing international meetings and encouraging the growth of brain research laboratories in developing countries by running training programs. While some nations formed interdisciplinary brain research organizations without influence from IBRO, in many cases there is a clear organizational lineage that shows IBRO as the catalyst that stimulated the formation of widely interdisciplinary brain research associations on a global basis. In the United States, for example, early IBRO members requested and served on the Committee on Brain Research of the National Academy of Sciences, established in 1964. Fostering brain research was a central concern for this well-connected group, and by 1968 it had done the spade work to start a U.S. association of neuroscientists.[6]

Before I turn to that development, however, I must mention another U.S. organization formed in the early 1960s. The Neurosciences Research Program (NRP) was the brain child of Francis O. Schmitt, a biologist at the Massachusetts Institute of Technology. Schmitt had been central in the institutional establishment of biophysics, as organizer and first chair of a new National Institutes of Health (NIH) Study Section on Biophysics and Biophysical Chemistry.[7] In the early 1960s, he became excited about the possibility of exploring the connections between molecular level interactions and brain processes such as memory. He began to gather around him a lively, informal discussion group, then to consider how to organize the effort on a longer term basis. "The Mens Project" was officially formed by a group of twelve men meeting in New York City on 1 February 1962. Schmitt moved quickly after this meeting to secure funding, submitting a grant application to NIH on March 1. A site visit team came to MIT five days later, and on March 14 the grant was approved. The title "Neurosciences Research Program" emerged in a grant application to The National Academy of Sciences (NAS) later that spring.[8] The NRP Associates were a select group, who met regularly for intense and stimulating discussions on the molecular basis of brain functioning. The NRP was highly influential in developing research problems for the field, particularly through a series of month-long Intensive Study Programs held at Boulder, Colorado from the late 1960s. Promising young researchers were invited to these study sessions, and absorbed the world view of the NRP Associates through them. The volumes of edited lectures from these series also served as the early textbooks for the field.

Schmitt, however, did not play a large personal role in the formation of what was to become the major institutional representation of an integrated interdisciplinary study of the brain, the Society for Neuroscience. One of the activities of the NAS Committee had been to collect the U.S. data for a world inventory of brain research manpower and facilities for IBRO.[9] The survey identified a surprisingly large number of local groups meeting around the country to explore the brain across disciplinary boundaries. The Committee recognized the possibility of forming a national association to bring these people together. The Society was officially formed in 1969, and held its first meeting in 1971, with 269 papers presented to about 1400 registrants. It expanded rapidly. By 1975, there were 1257 abstracts submitted and 3800 registrants. The Society itself by then had 43 local chapters and over 3900 members. Its 1994 meeting had over 20,000 registrants.

For all these professional organizations, the availability of funds was critical.[10] IBRO spent a great deal of early energy securing and re-securing funds from UNESCO for its work; its major competitor there was molecular biology. It went through a crisis in the early 1970s, when UNESCO funds were cut off, and it reorganized itself to include national members so that countries that could raise a few thousand dollars could finance the central office and newsletter. The U.S. Society for Neuroscience even contributed to this fund in its early years, before it was on a strong footing financially itself. Schmitt's first grant from the National Institutes of Health, mentioned above, was critical in getting his Neurosciences Research Program moving. Office space, staff, and travel funds for NRP Associates to attend the intense seminars that were its lifeblood demanded a constant effort to secure funds from many sources: primarily government agencies, but also some private foundations, and certainly appeals directly to wealthy individuals. The Society for Neuroscience owed its existence in the early years to grants from the Sloan Foundation, and the Grass, Markle, and Grant Foundations also contributed. Both IBRO and the Society for Neuroscience struggled over the question of whether to approach drug companies for support, and both decided in the end to create a special category of 'sustaining members' to accept drug company funds. In neither case is there any indication either that these funds came with any restrictions, or that they amounted to significant amounts except in comparison with the tiny treasuries of the organizations at the time.

Clearly, then, funding organizations need to be placed in the institutional map of the development of the neurosciences. In the co-production process, while funding organizations were helping neuroscientists to organize professionally, the professional organizations, and in particularly the spectacular early growth of the Society for Neuroscience, also encouraged the existence of funding programs. The early support for the field had come either from general sources or from problem-oriented programs to which neuroscience contributed. In the 1970s, however, basic neuroscience gained a foothold. An attempt in 1966 to establish a neurobiology program at the National Science Foundation had been blocked by the head of the physiological processes section, but a second attempt, bolstered with the growth statistics the new Society was proudly producing, could not be turned back on such grounds, and the program was set up in 1971. A Fundamental Neurosciences Program also appeared at NIH (in the Neurology Institute) in the mid-1970s, just at the height of the push for programmatic (not fundamental) research across the government.[11]

Standing behind the funding programs, however, is a broader set of political networks across government, networks which extend to Congress and the organized public. By 1978, members of the Society for Neuroscience were complaining of shortages of funds, and one of its presidents organized a symposium on federal support for the field.[12] This effort grew into a regular set of activities to, as this tax-exempt organization liked to put it, "educate members of Congress" on the excitement of neuroscience research frontiers. As the professional neuroscience lobby gained strength, it also gained allies in the form of private organizations seeking help for patients with particular diseases. Among the most powerful of these are the Alzheimer's advocacy groups, which formed a conscious alliance with the new National Institutes on Aging in the late 1970s to promote Alzheimer's research.[13] Many of these groups have 'research committees', chaired by prominent researchers, which form the site for negotiations over research agenda and political support. Historically in the United States, such coalitions of organized public groups, members of Congress, and biomedical researchers have been highly effective in increasing funds for research. The Congressional declaration of a "Decade of the Brain" in the 1990s indicates the power of this coalition in the neurosciences — though constrained budgets and still more effective lobbying by the Breast Cancer Coalition and AIDS activists has meant that not much money has followed the fine title.

As though this institutional picture were not already complex enough, I must mention one last category in the emerging organizational structure of the neurosciences: university departments. A few neuroscience departments appeared in the early rush of enthusiasm for the field, for example, at Harvard's Massachusetts General Hospital, at the new University of California at San Diego, and at the University of Florida School of Medicine. On most campuses, however, entrenched departments were not about to be displaced by youthful enthusiasms, no matter how impressive the growth numbers for the professional society. Through the 1970s, then, most researchers in the field were either trained in other fields or trained in interdisciplinary programs and institutes.[14] In the 1980s, however, enough external legitimacy seems to have been achieved, and the young neuroscientists of the 1970s had moved into high enough positions, that the rate of department formation picked up. It seems to be increasing even further in the 1990s.[15] Interestingly, however, these departments are often formed to keep a particular individual on a particular campus, rather than as an internal organizational need. Therefore, they often co-exist with neuroscience training programs on the same campus, making the institutional picture exceedingly complex.

WHY STUDY THE RESEARCH AGENDA?

By now, the historians among my readers must be reeling with the size of the historical task implied by the institutional inventory I have just provided. The NRP records, donated to the MIT archives when the organization moved to New York City, occupy 23 linear feet; luckily, only the first few boxes were available to me because of time restrictions. The IBRO archives are scattered, with several personal collections at the University of California Brain Research Archive in Los Angeles, some at the IBRO office in Paris, and most still in private hands.[16] Perhaps blessedly, the Society for Neuroscience has never organized its archives, which are in storage away from the main office, and although the Society has graciously made its early newsletter available to me,[17] its Council has declined to give me permission to look at Council meeting minutes even from the 1970s. Many of the major figures in the story of these organizations are still alive, and I have in fact appeared with some on history panels at meetings. The oral history possibilities are staggering in a field this large. Finally, the preceding discussion has not even touched on the problems and prospects of archival and oral history work on the funding organizations and non-governmental organizations, which will undoubtedly pose their own challenges.

Given this richness of resources for institutional history, why am I concerned about the research agenda at all? The answer comes from the theoretical issues my project set out to discuss. This project is not a conventional history, although I have used historical tools. Instead, it is an attempt to discover (as though that were possible) the source of the power of the neurosciences. How were they able to establish themselves in the funding system in the 1970s? What was the force that produced further dramatic growth in the 1980s, and the eventual result that in the early 1990s, the neurosciences were one of the few fields of American science still recruiting large numbers of American students? In formulating my project, I phrased this issue as the relationship between autonomy and power: Why didn't the growing political alliances bring growing pressure toward problem solving? How did the neurosciences gain autonomy in the form of professional organizations, basic research funding programs, and university departments?

The standard account of this phenomenon, which has been repeated among neuroscientists so many times that they have trouble considering alternatives, is that the field grew because of its intellectual excitement. A cultural object that my students and I have begun to call 'the romance of the brain' — the idea that the brain is a last frontier — animates much discussion around the growth and importance of the field. As we traced the influence of the neurosciences into the popular literature, however, we found an interesting phenomenon. The 'romance of the brain' seems to appeal primarily to people who are identified as 'brainy' in American society. Popular coverage of brain research findings is concentrated in high-brow (if you will excuse the pun) media, indicating that people whose identity is most tied up with their alleged mental abilities are also most likely to take an interest in how the brain functions.[18] The 'excitement' of brain research, seen in this context, is in part a sort of narcissism, an interest in one's self. This interpretation supplements the claim of objectively-identifiable 'excitement' that is usually invoked to justify increased resources for the field.

Further, we found that neuroscience knowledge has effects in the everyday lives of Americans in two forms: setting a framework for the understanding of conditions that are labeled diseases or disorders of the brain or nervous system; and incorporation into drugs and other therapies to treat these conditions. On the one hand, these influences shape the self-image of Americans, as more or less in control of themselves. On the other hand, they shape the actual life chances of Americans, as they are institutionalized or not, treated or not, diagnosed or not, in a co-produced relationship with available drugs and treatments. The details of this analysis are still to be filled in, but its importance is clear. Neuroscience is not just an autonomous field of science. It is also a force, if an invisible one to most observers, in the lives of average Americans.

It is particularly in this light that understanding the evolution of the research agenda of the field becomes important. If neuroscience knowledge is incorporated into the daily lives of millions of people on an ongoing basis, then it is important to know why it develops one way and *not* another. What alternative paths were not selected in this field as it evolved? The selection mechanisms have been described above: professional associations, funding bodies. To know what the alternatives were, however, we need to have a grasp of the research agenda.

VISUALIZING TOOLS

I have been particularly concerned with this question because in my early professional training, I was steeped in one of the major visualizing tools of contemporary history, bibliometric mapping techniques. For the first four years of my professional life I worked at the Institute for Scientific Information, the place that produces the *Science Citation Index*, and the home of co-citation clustering. Co-citation clustering is an algorithmic approach to analysis of relationships among published documents. Henry Small, the person I worked with there, has developed sophisticated techniques for identifying small, coherent research fronts in science and displaying them and their relationships in two-dimensional space.[19] The maps can be done either cross-sectionally or longitudinally. In my early work on the discovery of the opiate receptor,[20] I was aware of such maps, and they have surely been a part of my conceptual framework ever since. When I say 'research agenda', I mentally call up such a map, and I can picture it changing over time.

In my work on the neurosciences, I have experimented with using this sort of data, and the array of related bibliometric techniques that have been developed since Small's early work. I have only frustration to report thus far, however, not results. Let me mention here, therefore, some of the possibilities and difficulties of making maps of the neurosciences over the period I am interested in.

Ideally, I would begin mapping the research agenda of the neurosciences with a baseline map for about 1960, the time when IBRO and the NRP were first starting. At this time, there were no interdisciplinary brain research journals, with the possible exception of *Experimental Neurology*, founded in 1959. Most of the work of the neuroscience pioneers was being published across the biomedical literature. One of my students has in fact produced the raw data for a baseline map by typing titles of articles written in the late 1950s by early members of IBRO and the NRP into a computer file. We will be producing a map using the words that appear in those titles with standard statistical techniques, including factor analysis and multi-dimensional scaling. From looking at the set of article titles, however, I already have a feeling for what the map will look like. These people were working with distinct tools in quite well-defined and very separate areas of the nervous system, on a mix of basic exploration of the nervous system and problem-oriented (mostly drug-related) topics. A great deal was known about the anatomy of the brain at that point, and physiological experimentation was pulling the anatomical knowledge into a common framework of knowledge about processes, but there was no complex web of inter-related explorations at all comparable to the inter-connectedness of the contemporary neurosciences. The pioneers were spread rather thinly over what we now perceive to be a large territory.

If I wanted a second map for 1970, after the NRP had been in operation and held its first study program, and just at the point when the Society for Neuroscience was forming, I would have a different set of tools available. Two interdisciplinary journals had been founded, *Experimental Brain Research* and *Brain Research*, and are included in on-line bibliographic data bases. We have in fact down-loaded this data, and are working on analyses of these journals, through the same statistical approaches as before, but without the laborious hand work of entering article titles. This strategy loses the information from articles published outside these two core journals, and we can assume that at this early stage of development there were still a large number of those. It has the advantage, however, of not relying entirely on lists of organizational activists or elites to identify data; the activities of a broader set of neuroscientists can presumably be included. In addition, the co-citation cluster archive at the Institute for Scientific Information actually starts with 1970 data; but it is extremely awkward to find the neuroscience clusters in those early printouts, which are not well indexed, so limited information is available. The tapes originally used to produce the printouts have long since been erased.

The 1980 map would be a different story. By this time, a number of the professional associations themselves had started journals, and independent ones had formed as well: *Neuroscience Letters* (1975), *Neuroscience* (1976), *Brain Research Bulletin* (1976). *The Journal of Neuroscience* followed soon afterward (1981). The data is available on-line. The ISI cluster archive is better indexed, and tapes exist to inventory the clusters using the core neuroscience journals, a much more efficient way to search for neuroscience-related clusters than searching by hand through the dusty 1970 printouts. However, along with convenience comes information overload. In my project, I actually had ISI run a cluster inventory for several years in the early 1980s. Using a small set of core neuroscience journals as tracers, nearly a quarter of the entire cluster archive was identified as neuroscience-related, on the order of 1100–1300 clusters out of approximately 4000 produced each year. The work of producing maps from such a mass of data would be a project in itself, and we have not yet attempted it in the context of my current modest grant. The co-citation technique has also been used to link clusters over time, and a wealth of time-series data is available at ISI for both the 1970s and 1980s.[21] Again, however, a major investment of time would be required to locate, sort, and diagram that data in any way that provided an overview.[22]

WHY MAP?

In principle, then, maps of the research agenda of the neurosciences could be produced for these various points in time, with greater or lesser methodological sophistication, using various sets of assumptions. Determining from them what possibilities in the field had not been explored is possible, although the discontinuities in data series present something of an obstacle. The effort involved in doing this, however, immediately raises the question of payoff. What would we know after all this effort that we did not know before, or could not find out using other approaches?

With regard to the second part of this question, I have started asking past presidents of the Society for Neuroscience, who I am interviewing on organizational history, to tell me what they see as the major developments in the field in the 1970s, the period when they were active and served as presidents. I have only two cases to report, but the difference in their reactions is instructive. One past president drew an absolute blank on this question, and eventually responded by talking about his own research area. He had clearly either never had a good mental map of the field and its directions, or had thoroughly forgotten what was on it in the 1970s. The other past president, however, had an immediate and articulate answer. This second person was someone who had always taken an interest in an overview of the field, and had also been active in Congressional testimony and science policy discussions about resources for neuroscience. Therefore, through his interactive network, this person had a reason for maintaining a plausible and fairly simple mental map of the field.

The interviews point to the question of the ontological status of bibliometric maps, and their relationship to the overall analysis I am attempting of neuroscience knowledge as it is co-produced with politics, culture, and social organization. The maps are supposed to be indicators of the research agenda of the field. Clearly, 'research agenda' is a construct. Whose construct? Under what circumstances is it used? With what results?

Let me speculate a bit based on personal experience. I suspect that 'research agenda' is not a construct used by working neuroscientists, except in relation to their own particular work. Most neuroscientists are more likely to visualize their own research areas in terms of the other laboratories working on their problems than in terms of abstract citation relationships.[23] When they turn to the problem of understanding the structure of the field of neuroscience, I suspect that most working neuroscientists would use a different sort of mental map, passed on to them in the textbooks in the field and introductory courses. This map is basically organized as a matrix. One axis of this map is anatomical, reflecting areas of the brain. The other axis is methodological, covering categories like physiology, anatomy, and now molecular biology, since any of these approaches can be applied within any brain area to discover the specifics of the various brain processes. There is a supplementary diagram to this map, a kind of inset, that includes detail from the cellular level down through chemical and molecular components of the nerve, i.e., the research areas where the presently-labeled 'fundamental' research of neuroscience is taking place. This category scheme is probably sufficient for most working neuroscientists to place themselves mentally in relation to such concrete choices as, for example, which sessions to attend at professional meetings or which literatures to scan outside their specialty areas. It bears no resemblance, however, to the kinds of research-front macro-maps I described in the last section.

In fact, I suspect that the kind of macro-map I have described exists in the minds of only a limited number of neuroscientists, including some senior heads of laboratories and well-used advisors to government like the past Society for Neuroscience president I interviewed, who are forced by these responsibilities to develop the overview. It is these researchers who can see the alternative paths the field might follow, and who therefore can make choices for the field in some form. Unless one can see that fuller landscape, day to day decisions about research directions are more constrained than liberated. This is probably the situation of most working neuroscientists.

If I am right in my speculations, 'research agenda' is a recognizable construct for some leaders of the neuroscience research community. Clearly, however, it is also *my* construct. It has played a role in the formulation of my project, and I continue to feel the need for maps to make sense of the seamless web of co-produced social reality of which neuroscience is a part. Why? In part, of course, I need them because of my disciplinary commitments: because sociologists of science can no longer leave the content of scientific knowledge out of their conceptual schemes. Another reason, however, arises from the goals and emerging issues of my particular project. While the data that go into the map-making exercise come from the relatively isolated actions of laboratories that do not necessarily see its overall contours, it is the *aggregate* tendencies that appear in the map itself that produce the *patterns* of consequences for everyday life that are ultimately of concern to me in this research. One example is the reductionism that creeps into the self-image of patients when doctors view their every complaint as an opportunity for drug intervention. Another is the differential access to expensive treatments that may leave poor people institutionalized or alone on the streets while those with more money lead normal lives with their families.

Without the construct of the research agenda to link laboratories through knowledge to such consequences, we cannot see the full context of neuroscience, and therefore cannot open the question of choice to the people who pay for and live with the research.

It may be that I do not need the actual maps to accomplish that result. Perhaps all I need is the construct of the research agenda, and access to the maps already being drawn by leaders of the neuroscience community, to be able to write intelligibly about the web of relations that link research and its consequences. Whether the maps become a reality or remain a construct, however, they certainly make me part of the multiple-map phenomenon that constitutes the contemporary history of science.

ACKNOWLEDGEMENTS

I gratefully acknowledge the creative and energetic help of several graduate students in the course of this project: Kate Lytton, Diane Shichtman, Shana Solomon, and Xuemei Xie.

NOTES

1 I borrow this term from Sheila Jasanoff.
2 Latour's (1987) suggestion. I apologize for gendering my research object here, but I am forced to do so by the research strategy I am describing, which treats singular scientists, rather than plural ones who can be discussed gender-neutrally.
3 Again, using Latour's term, which refers to the seamless web of science, politics, social relations, and culture woven by interaction.
4 Frank, Marshall, and Magoun 1976; Magoun 1962.
5 Cozzens 1996.
6 Frank, Marshall, and Magoun 1976, 604.
7 Swazey 1975, 530.
8 Swazey 1975, 535–537.
9 Frank, Marshall, and Magoun 1976, 605.
10 Here I begin to draw on my archival work at the NRP Archives at MIT, IBRO archives at UCLA, and copies of the newsletter of the Society for Neuroscience in the 1970s.
11 With the exception of a few records at The National Science Foundation (copies in my files), I have not yet begun to explore the archives or gather oral histories of these programs, although they are a part of my research plan.
12 *Neuroscience Newsletter* vol. 9 (number 4, December 1978), 1.
13 Fox 1989.

14 Department and program figures from the series Neuroscience Training Programs of North America, produced by the Society for Neuroscience.

15 Susan E. Cozzens, Diane Shichtman, and Shana Solomon, "Neuroscience departments: the search for autonomy", paper given at the meeting of the International Society for History, Philosophy, and Social Studies of Biology, Brandeis, MA, July 1992.

16 Again, I have seen and worked with many linear feet, and still not had the energy after several trips to UCLA to delve into two large file drawers of complete correspondence of two early secretary-generals.

17 In early interviews with presidents from that period, the coverage of issues in the newsletters seems to be validated as fairly complete.

18 Susan E. Cozzens, Kate Lytton, and Diane Shichtman, "Women, men, and neuroscience", paper presented at the meeting of the International Society for History, Philosophy, and Social Studies of Biology, Evanston, IL, July 1991.

19 For an overview of co-citation clustering at ISI, see Small and Garfield 1985.

20 Cozzens 1989.

21 The two use somewhat different methods, however, so that the decades cannot be linked very persuasively.

22 In another project, I used journal-journal citation techniques to produce maps of the field at the journal level. These display interesting relationships with related broad areas of biomedicine, but by no means reveal the research agenda of the field at an appropriate level of detail to answer my questions.

23 Past experience has shown, though, that there is a fairly good relationship between the contributors located on co-citation maps and the sets of contributors researchers in the field recognize — not surprisingly, since those researchers created the citation relationships that were the raw data for the maps.

BIBLIOGRAPHY

Cozzens, Susan E., *Social Control and Multiple Discovery in Science: The Opiate Receptor Case* (Albany, NY: SUNY Press, 1989).

Cozzens, Susan, E., "IBRO in national and international perspectives", *Neuroscience*, vol. 72 (1996), 301–306.

Fox, Patrick, "From senility to alzheimer's disease: the rise of the Alzheimer's disease movement", *The Milbank Quarterly*, vol. 67 (1989), 58–102.

Frank, Robert J., Louise H. Marshall, and H.W. Magoun, "The neurosciences", in John Z. Bowers and Elizabeth F. Purcell (eds.), *Advances in American Medicine: Essays at the Bicentennial*, vol. 2 (New York: Josiah Macy, Jr. Foundation, 1976).

Magoun, Horace W., "Development of brain research institutes", in John D. French (ed.) *Frontiers in Brain Research* (New York: Columbia University Press, 1962).

Latour, Bruno, *Science in Action* (Milton Keynes: Open University Press, 1987).
Small, H. and E. Garfield, "The geography of science: disciplinary and national mappings", *Journal of Information Science*, vol. 11 (1985), 147–159.
Swazey, Judith P., "On the Nature of Research in the Neurosciences", in Frederic G. Worden, Judith P. Swazey and George Adelman (eds.): *The Neurosciences: Paths of Discovery*, vol. 1 (Boston: Birkhaüser, 1992).

CHAPTER 10

Writing about Scientists of the Near Past

Frederic L. Holmes

INTRODUCTION: THE PARADOX FOR RETROSPECTIVE HINDSIGHT

In his general overview of the problems confronting historians of contemporary science, Thomas Söderqvist stresses both the desirability of writing the history of recent science and some of the formidable obstacles to its achievement.[1] The issues raised are so important that I would like to frame what I have to say here in large part as a response to the positions he has taken. My primary qualification for doing so is that I have scattered my own historical efforts in several eras, ranging from the seventeenth century to the 1950s. My experiences dealing with problems of documentation, scale, and perspective in these diverse periods can, I hope, provide some insight into the nature and seriousness of the problems that Söderqvist foresees for those writing about contemporary, or nearly contemporary science.

To begin, we should distinguish between two types of problem raised. One set of Söderqvist's concerns, relating to the lack of temporal perspective on events happening in the historian's own time, we share with our own predecessors. Eighteenth century historians writing about the science of that century had no previous historians to rely on, were sometimes in a position to talk with their subjects, and lacked the perspective of hindsight to help them distinguish between transient and enduring results. So with the more numerous nineteenth century historians. That did not prevent them from writing much which is still of great value to historians of science of today, who unfortunately return too seldom to the guidance those works can still provide. Another set of Söderqvist's concerns revolves around factors that we commonly believe to be unique to science since 1945: i.e., its vastness, its technical complexity, and the dawn of the computer age which has so altered the nature of communication and of the documentary record.

One of the curious aspects of the first problem is that we worry about the lack of retrospective insight when we think about writing contemporary history, but when we avail ourselves of such insight to write about the deeper past, we worry equally that we are being Whiggish or presentist. We claim that we should immerse ourselves in the science of the past, to evaluate scientists or the scientific work of any earlier era in its own terms, according to the standards and the general state of knowledge of that time, not according to later knowledge and standards. But when we write about events so recent that we have no later standards to apply, we view their absence as a liability.

What are we to make of this paradox? The pessimistic view would be that, if both hindsight and lack of hindsight interfere with a clear view of our subject, then the historiography of both contemporary and older science is inevitably problematic. A more optimistic view is that these reciprocal problems can be used to advantage; that experiences working in the near and far past can be complementary, that the pitfalls encountered in one can help us avoid the pitfalls of the other; that the long vistas through which we view events from which we are temporally distant can help us to gain perspective on events that still crowd in upon us; and that the more intimate connections we can establish with contemporary events and their participants can help us move closer in spirit to those who lived and acted far before us.

Treating the science of former times in its own context is very different from treating contemporary science in its current context, because, as inhabitants of the present, historians must make concerted efforts to enter the mental world of scientists of the past. When dealing with contemporary science, on the other hand, we must make a concerted effort to escape from a mental world we may share with our subjects, if we are to view them with appropriate detachment. I will argue that it is easier to achieve the latter, if we have some experience with the former. In other words, an effective way to prepare for writing about the history of contemporary science is to have written about earlier science.

This point can be clearly illustrated from the case of eighteenth century, pre-Lavoisier chemistry. Reading the chemical texts of this era is not easy. The language is foreign, the details are technical, modes of reasoning and criteria of composition on which the modern chemist relies are missing. Historians of chemistry habitually try to make access easier by translating the original descriptions of chemical operations into modern formulas and equations. This device may make it easier for the introductory reader to bring the remote closer to the familiar, but actually hinders us from entering the world of the eighteenth century chemist. If one reads extensively in the original literature, such translation eventually becomes unnecessary. Just as an Anglophone learns gradually to read French or German without translating, so we learn to follow the thoughts and operations of eighteenth century chemists in their own language, and to observe them confronting their problems in their own terms.

Recently I tried an experiment in a seminar I give on the chemical revolution. We read sections of Pierre Joseph Macquer's *Elements of the Theory and Prac-*

tice of Chymistry[2] as though we were students who were beginning to learn chemistry in the 1760s. I asked one student to lecture to us on what we needed to know to pass the course. I insisted that in discussing the text, no one should invoke language or ideas from later eras, including the present. My students entered so fully into the spirit of the exercise that when we later read Lavoisier they tended to resist his claims.

In this role-playing, of course, the students did not abandon their belief in modern chemistry, but they glimpsed what it was like to think within an era that no longer exists. As practicing historians we do so for much longer periods in our scholarly lives. That we can think in ways that we know belonged to eras of which we are not permanent members can, I believe, help us as historians to think about contemporary science. The science of our day, too, will someday belong to an earlier era. Merely to state that truism is, however, not enough to teach us how to deal with it. The experience of dealing with older science can lead us to a deeper understanding of what it means to enter into the thought of our current age without being captive to it.

IS THE SCALE AND TECHNICAL DIFFICULTY OF RECENT SCIENCE SO DIFFERENT?

I will devote the rest of my time to Söderqvist's assertion that the study of science since 1945 confronts the historian with "new and unfamiliar problems" due to the immense volume and technical difficulty of recent scientific work. The volume of research, the flood of documents, and the conceptual and methodological complexity of contemporary science are, indeed, daunting. I will argue, however, that such problems are already formidable for the historian of science of any period since at least the early nineteenth century — i.e., since the era in which science became organized within disciplines represented by academic specialists who were expected to pursue research as a matter of course.

There may be millions of research reports per year in the post-1945 period, compared to only thousands per year in the mid- and late nineteenth century; but thousands per year is already enough to overwhelm historical analysis. The 'information explosion' that created a need for abstracts and reviews of the current scientific literature was already visible to scientists in several specialties by the 1830s. In that period Jöns Berzelius began putting out a *Jahresbericht* in which he summarized and evaluated the more important papers published in chemistry during the preceding year. Disciplinary journals such as Johannes Müller's *Archiv für Anatomie und Physiologie* also began to publish annual reviews of developments in these fields. The problem of keeping up with one's field, about which one hears so much in our time, was already acute for scientists in the active, growing fields of 150 years ago.

The problem of producing a grand historical narrative of the development of one of these nineteenth century specialties is, I believe, nearly as formidable as for the specialties of the post-1945 era. As Söderqvist has noted, Derek Price asserted in the 1960s that science had for more than 300 years grown exponentially, with the "remarkable rapidity of a factor of 10 in 50 years". The total output of science in the post-1945 era must, therefore, if Price was right, be something more than 1000 times that of the mid-nineteenth century. That does not mean, however, that each contemporary scientific discipline is 1000 times as large as a nineteenth century discipline. Price also argued that a scientific field has typically contained "up to a few hundred" actively productive scientists, because of factors such as the limit on the number who can "monitor" one another's work. "When in the course of natural growth it begins sensibly to exceed" this number, the field tends to divide into subfields.[3]

Price made his case deductively, rather than by gathering historical data on the size of specific fields over successive scientific generations; but the plausibility of his argument suggests that the 'vastness' of the science of our time compared to that of earlier times is to be measured in the number of specialty fields, rather than the size of each such field. At the beginning of the nineteenth century there were only a few sciences recognized as disciplines — for example, physics, chemistry, geology, anatomy and physiology, botany, and zoology. The emergence during the 1830s of inorganic and organic chemistry as distinct subfields may be the earliest modern example of the dynamic postulated by Price.

If this is the basic pattern of growth in modern science, then historians of contemporary science do face problems of unprecedented scale in attempting to map the entire world of science. In attempting to depict the development of particular disciplines, however, I would suggest that their problems are similar in magnitude to those studying the science of the early twentieth century and most of the nineteenth century.

"The technical difficulty of recent science" to which Söderqvist addresses justifiable concern, does not distinguish recent science so sharply from that of earlier periods as he has suggested. If we are less aware of the technical difficulty of older science than we should be, that is probably in the main because, except for mathematics, physics, and astronomy, historians have tended to avoid the more difficult regions within the sciences of the past. As a historian of the life sciences, I want to respond particularly to Söderqvist's statement that:

> "the recent life sciences stand out as technically more difficult to understand today, because they have become much less descriptive, much less phenomenal, much less common-sensical, and much more abstract and mathematized since 1945, partly as a result of the rapid infusion of the physical and chemical sciences in biology and medicine".[4]

All of the trends mentioned here have been in motion since long before 1945. It is probably meaningless to seek a beginning for these movements, but the statement would apply with special force to the historical development of the life sciences if we were to substitute the date 1820 for 1945.

In 1822 the French Academy of Sciences offered a prize for an experimental investigation of the chemical processes of digestion. The justification given was that the more precise new methods of chemistry provided new opportunities to elucidate a formerly mysterious physiological process.[5] This announcement can be taken historically as a symbol of the opening of an era of rapid infusion of chemical methods into the life sciences. The methods in question were not the glamorous new quantitative ones we associate with the chemical revolution, but an expanding repertoire of qualitative, yet technically demanding, procedures for extracting, isolating, and characterizing the distinctive substances making up the solids and fluids of plant and animal bodies. The successes achieved in the application of these methods to the complexities of the internal processes of digestion, nutrition, secretion, and so on were not at first dramatic, but they were cumulative, and they were starting points for investigative traditions that have continued into our own century. Anyone who thinks that the early stages of this investigative movement were technically simple should read through the ninth volume of Jöns Jacob Berzelius's *Lehrbuch der Chemie*, published in 1840, which summarizes the progress of what was then known as animal chemistry up until that year.[6] Here is a good test of the historian's grasp of these technical difficulties: in 1813, Berzelius reported that the three known substances characteristic of blood — fibrin, albumin, and the "coloring matter" — "resemble each other so closely that they can be considered modifications of one single substance".[7] By 1839, it was possible to separate the "coloring matter" into two components, one of which — named hematin — was responsible for the red color, while the colorless component could be identified with confidence as different from either fibrin or albumin. Berzelius called this new substance globulin.[8] The fundamental modes of analysis that enabled these new distinctions to be made were similar to those that Berzelius was already applying in 1813. How much technical competence do we need as historians, to explain why physiological chemists could achieve in the late 1830s something that had been 20 years earlier still beyond the reach of the leading practitioner of the methods used?

Recently I have begun to explore a topic that will, I believe, illustrate to advantage the infusion of physics into physiology during this same period: i.e., in the study of the mechanics of the circulation of the blood. Such investigations were not new to the nineteenth century — we are all familiar with Stephen Hales' celebrated treatise on *Hemostatics* — but they were taken up in the 1820s with a new degree of technical depth, by scientists well-trained in physics and mathematics, or in collaboration with others who were. Two figures dominated the early phase of this development, Jean Léonard Marie Poiseuille in France, and Ernst Heinrich Weber in Germany. In 1828 Poiseuille, a former student of the *Ecole polytechnique*, measured "the force with which the heart drives blood into the aorta" for his doctoral dissertation in medicine. Using a specially designed "hemodynamometer", he performed experiments that had, according to a contemporary reviewer, a "character of precision which commands confidence in their results, even if one is a foreigner to mathematics and cannot follow the author through his calculations". Contrary to his own expectations, Poiseuille found that the average force calculated from a long series of measurements was the same in arteries at different distances from the heart. He also confirmed that the average force decreased during respiratory inspirations and increased during expirations. These and other regularities that the sensitivity and precision of Poiseuille's results revealed opened up questions about the dynamics of the circulation that demanded further precise measurements and informed theoretical explanations.[9]

During the 1820s Ernst Heinrich Weber, an anatomist, collaborated with his brother, the physicist Wilhelm Weber, in a comprehensive experimental investigation of wave motion. To observe the waves that spread through fluids, they constructed long, narrow troughs, viewing the motions through the glass sides of their apparatus. They also studied waves produced by fluids forced through closed tubes. From these studies they were able, among other things, to make a clear distinction between the motions of the wave forms and of the particles comprising the fluids through which the waves moved.[10] Afterward, Ernst Heinrich Weber applied these results to reinterpret the motions of the blood through the arteries, and the arterial pulse. The wave motion which the force of the blood caused to pass along the elastic walls of the arteries was, Weber emphasized, distinct from, and much more rapid than, the actual movement of the blood through the arteries. It was also the elasticity of the walls that converted the pulsatile motion of the heart into a continuous motion of the blood within the arteries.[11]

In applying hydrodynamics to the problem of the circulation of the blood, Poiseuille, Weber, and others deployed technically difficult experimental and theoretical methods, drawn from physics, to a central problem in physiology. Some of the physiologists who took up these problems were not trained in physics and mathematics. Hermann Helmholtz and Emil Du Bois-Reymond, who did know these subjects well, criticized the *Hemodynamics* of the capable physiological experimentalist Alfred Volkmann, on the grounds that Volkmann did not understand the ABCs of physical theory.[12] How can historians of science adjudicate such opinions, or more generally interpret developments in this field, unless they are more competent in these matters than was Volkmann himself?

It would be easy to multiply such examples to show that the problem of the basic training of historians of science in science, which is, according to Söderqvist, "acute, serious and seemingly unavoidable" for historians of recent science,[13] ought to be equally so for historians of older science. I now want to turn to an aspect of recent science which I believe does present problems that, although similar in kind to earlier science, manifest themselves with greater intensity the closer we come to the present. These problems relate to the ever-increasing overall quantity of scientific activity.

One of the prerequisites for a history of science adequate to its task, is that there be a sufficient number of scholars in each of its own sub-fields to provide critical safeguards, to build broad conclusions from collective effort, to cover the many layers of investigation and interpretation necessary to link densely specialized monographs with broader syntheses. No matter how careful and rigorous the individual scholar may be, she cannot encompass all of the relevant skills, possess all of the pertinent viewpoints, or obtain all of the cogent evidence to narrate and interpret alone any complex historical episode. Like any discipline, ours progresses most effectively where there are enough people engaged in overlapping studies to have read some of the same sources, to complement or to challenge one another's conclusions, to catch the omission of a key piece of evidence or the questionable interpretation of another.

Historians of science often point with pride to the rapid growth of our field during the last three decades. We must not conceal from ourselves, however, that, relative to the vast intellectual terrain for which we hold scholarly responsibility, we remain thinly scattered settlers. We have established a few well-populated strongholds, beyond which we can claim only widely dispersed frontier outposts.

Because our predecessors of the immediate post-war generation focused their efforts on the events we summarize under the phrase 'scientific revolution', we have attained reasonable concentrations of scholarship around the major scientific figures and events from Copernicus to Newton. As we ventured into the nineteenth century and beyond, however, we have increasingly fanned out into uncharted territories. Perhaps in the field of theoretical physics, which possesses a strong master narrative into the twentieth century, there exists a critical mass of scholarship reaching almost to contemporary science. The situation is not that advanced within the fields with which I am more familiar.

In the history of chemistry a strong cluster of scholarship continues to surround Lavoisier and the chemical revolution. There has been, periodically, sufficient attention given to John Dalton and the atomic theory to stimulate critical debate over sources and interpretations. Beyond that, it is hard to find an individual scientist or historical set of events on which more than a handful of historians are engaged in comprehensive studies. In the life sciences we can be confident that Darwinian evolution will continue to attract younger scholars to sustain what has been for several decades an extremely lively focus of study and debate. Gregor Mendel continues to fascinate enough historians to pore repeatedly over the sparse documentation that has survived about his life and work, and to promote critical reevaluation. Vast reaches of the life sciences, however, including particularly the experimental life sciences, remain underexplored. The irony of Söderqvist's quote, "Who will sort out the hundred or more Paul Ehrlichs?"[14] is that even the single historical figure of Paul Ehrlich has not been fully sorted out. Despite a relatively rich collection of manuscript sources, now deposited at the Rockefeller archives, Ehrlich has still not received the concentrated scholarship necessary to transcend his legend, to interpret his scientific life, or to assess critically his place within the broader historical developments to which he made such highly visible contributions.

I do not want to present a generally negative view of historical scholarship in the modern sciences. There are some very bright spots in the landscapes I have mentioned. Impressive individual works have recently appeared on nineteenth and twentieth century chemistry, twentieth century genetics, and other areas of the life sciences. But as these authors would probably attest, a historian can feel rather lonely in such endeavors. There are satisfactions in opening up new vistas, in having at one's disposal stores of previously unexamined documents, in providing the first detailed interpretations of highly significant, creative scientists. Those of us who have enjoyed such experiences are, however, keenly aware of the limitations of what one can achieve in territories that are too vast to survey fully and accurately by oneself. We are aware also, of the extent to which interpretations we make that we know to be debateable, partial, or provisional, pass unchallenged, because no one else has been over the same ground.

Nothing is more common, of course, than for each of us who cultivates some particular corner of these large stretches of landscape to call for more scholarship in the domains surrounding his or her own work. The problem is that there are not now, and probably never will be, enough productive historians of science to go around. Our era of rapid growth, like that of most academic disciplines, is largely in the past. Although there may continue to be modest increases in our numbers, in response to a spreading consciousness of the extent to which science and technology dominate the present world, an international academic community, for example, itself under constant pressure to conserve rather than to expand its resources cannot, in the foreseeable future, provide more than that.

The consequences of this predictable permanent shortage of historians of science are most dire for our efforts to cope with contemporary and recent science, where the disproportion between the numbers of active scientists and the numbers of historians of science able to write about them is most severe. Even if all of the 90 percent or so of historians that Söderqvist estimates to be working in periods before 1945 were to move into the contemporary era, we would still face a huge imbalance between the amount of science produced and the amount of probing scholarship about that science that we can compile. Is there anything we can do to increase the odds that we will deploy our relatively slim resources effectively?

When a small force is required to occupy a large territory, the most obvious question to be resolved is whether it is best to spread out evenly over it, or to concentrate on the control of certain strategic points within it. The analogy to our situation is easy to grasp. Should we try our best to include all of the fields and the major subfields, periods and events, of recent science in our scholarly enterprise, or should we group ourselves instead around a few dominant or representative stories and let the others go for now?

In our field, as perhaps in all academic fields with similar ratios between potential areas of study and available scholarly output, the natural tendency seems to be to form clusters around certain highly visible problems or events, leaving relatively vacant spaces between them.

In the life sciences in the twentieth century, for example, there has been sustained focus on the 'modern synthesis' in evolutionary biology, relatively steady, but seldom intense, activity relating to the formation of classical genetics from 1900 to the 1930s, and a recent surge of interest oriented around the formation of molecular biology. There are a few additional pockets of activity in which at least several major individual projects have been completed or are underway, including the history of immunology. To judge from the topics of several recent conferences, sessions in annual meetings, and special issues of journals, however, the history of molecular biology is currently attracting far more attention than any other area of the contemporary experimental life sciences. A canonical set of events, largely those treated either in Robert Olby's *The Path to the Double Helix*,[15] or Howard Judson's *The Eighth Day of Creation*,[16] sets the boundaries within which a growing cluster of historians are exploring more intensively a multitude of sub-themes and issues, and digging more deeply into the roles of some of the leading personalities involved.

There is no mystery about the drawing power of this subject for historical examination. Molecular biology as a field is both glamorous and powerful. Its progeny, including genetic engineering, genetic counseling, and the genome project, reach beyond the laboratory far into the political, industrial, and ethical realms. Many of its large-scale implications for our present and future well-being can be traced as historical consequences emerging from the dramatic small-scale story of Watson, Crick, and the double helix. James Watson is a quintessential symbol for our times. Irreverent, brash, awkward, scientifically smart and politically shrewd, he can appear heroic and also mock the more straightforward heroes of classical history of science. Francis Crick is a figure of legendary proportions, and there is a supporting cast of vivid, strong-willed individuals, including particularly Jacques Monod and Max Delbrück. Trans-Atlantic collaborations, such as those between the Institut Pasteur and Delbrück's phage group at Caltech appear in retrospect as harbingers of the global international networks of contemporary science.

Moreover, the group who came to dominate the formative developments in this saga constructed its own account of how it had all come to pass: a story refracted particularly through the reminiscences published in 1966 in the volume of essays entitled *Phage and the Origins of Molecular Biology*.[17] The story is replete with dramatic tones, featuring a small band of pioneers dissatisfied with the traditional fields, who bore, beneath their informal, iconoclastic style, a grand new vision of how to uncover the 'secret of life'. Resisted by the leadership of such well-established fields as biochemistry, this initially compact network of innovative scientists created their own discipline and soon made it the foremost field in the contemporary life sciences. So complete has their hegemony become that other fields have been scrambling to rename their disciplines to include, however clumsily, the word 'molecular' in their titles. Such a story presents a formidable challenge for historians of science. All of the methodological problems enumerated in Söderqvist's paper apply forcefully to this case. Moreover, the powerful mythology created by the founders themselves can easily seduce unwary historians. To surmount such hurdles, to contend with subjects as technically demanding as X-ray crystallography, to resist the charisma of the early leaders of the movement, some of whom are still around to charm and sometimes to intimidate a young scholar, historians of science must attack this story, with its multiple subthemes and subplots, in force.

Promising beginnings have been made. Historians of science are already moving beyond the core myths of molecular biology as they subject its various facets to closer scrutiny. In a recent issue of the *Journal of the History of Biology* devoted to "Building Molecular Biology", for example, Richard Burian argues that molecular biology "is not a discipline, because it does not center on a focal group of questions".[18] In the same issue Hans-Jörg Rheinberger asserts that, contrary to the myth put forth by members of the club, "what we today call 'molecular biology' emerged from and was supported by a multiplicity of widely scattered, differently embedded, and loosely, if at all connected experimental systems for characterizing living beings to the level of biologically relevant macromolecules".[19] To solidify such claims, to construct an account of the many events making up a story even broader and more complex than that encompassed within the founders' mythology, will require as many historians of science as our field can muster.

In view of these considerations, why should we have any misgivings about the fact that the small band of historians of the contemporary life sciences appear to be gravitating preferentially toward the history of molecular biology? Is it not better to cover one story well than to touch many stories superficially? The advantages are obvious; but the disadvantages need more attention than they receive. Even while historians are reaching a critical mass in this field that may enhance both its scope and its critical levels of interpretation, they may be participating, at a broader level, in further distortions of the overall history of the contemporary life sciences. The high visibility of molecular biology will become further amplified in our historical accounts of this period, by the very fact of receiving historical attention at the expense of other fields. Those who are already prominent may be made more so by historians, while the less visible become historically invisible.

Let me illustrate my contention. One of the prominent features of the founders' version of the origins of molecular biology was that they had solved the problem of the gene by by-passing traditional biochemistry. Watson expressed their collective disdain for classical biochemists pungently in *The Double Helix*. He was bored in the laboratory of Herman Kalckar, Watson relates, "because I could not see how the type of problem on which he was then working (the metabolism of the nucleotides) would lead to anything of immediate interest to genetics". Elsewhere he wrote, "It was bad enough learning crystallography without having to acquire the witchcraft-like techniques of the biochemist".[20] These were not mere expressions of the personal idiosyncrasies of James Watson, but of attitudes shared within the group of phage biologists from which he came.

Historians such as Burian and Rheinberger are already showing that biochemists played larger roles in the framing of critical concepts within the structure of molecular biology than they have been credited with by founders of a field who defined themselves in part by distancing themselves from the biochemists. Nevertheless, their central historical subject continues to be the formation of molecular biology. Biochemists are being included within a revised version of a story from which their rivals had tended to exclude them, but they are not yet portrayed as central characters in their own stories.

In a trans-Atlantic historical collaboration which emulates that of their subjects, Angela Creager and Jean-Paul Gaudillière are elucidating the way in which studies of the feedback regulation of bacterial metabolic pathways carried out by biochemists during the 1950s were integrated by Monod and Jean-Pierre Changeaux into the concept of allosteric regulation. Creager and Gaudillière present the historical process as a shift away from the problems in metabolic regulation toward those of gene expression, and eventually to molecular structure. They are mapping out a coherent, persuasive narrative, which adds much to our understanding of the contributions of biochemists to the formation of the conceptual structure of molecular biology.[21] Their perception that there was a shift away from the early problems in metabolic regulation, however, is an artifact. The studies on bacterial metabolism that became incorporated into this development were only one segment of a broader movement within the field of intermediary metabolism toward the investigation of the regulation of the many metabolic pathways that had by then been identified. The study of metabolic regulation continued to expand during the 1960s and afterward, becoming a focal problem within a metabolic biochemistry that still thrives as one of the 'mature' fields within the contemporary life sciences.[22]

The lesson here is that the studies in feedback regulation that Creager and Gaudillière have described are parts of more than one story. They are being placed historically only in the story dominated by molecular biology, in which metabolic biochemistry is reduced to a subsidiary stream. The other story remains historically invisible, because historians have so far not been attracted to the history of intermediary metabolism.

What this example illustrates is that, whether we concentrate or disperse our historical resources, we will encounter many pitfalls, because we are so few in relation to the overall scope of our historical territory. If we try to cover its many regions evenly, we cannot amass enough scholarly strength in any one place to probe both deeply and widely, to provide a robust counterforce to the stories offered to us by our subjects, or to check one another's interpretations. If we move selectively into a few favored niches, we risk distorting the overall picture by further exaggerating the relative prominence of fields and persons who have already achieved high profiles by the time we begin to examine their historical places. The dilemmas we face due to the smallness of our numbers in proportion to the vast scale of the activity of those we seek to study are so deep, I believe, that they overshadow all of the many other problems we may pose for the writing of the history of contemporary science.

ACKNOWLEDGEMENTS

I wish to thank Angela Creager for sharing with me the early draft of her impressive study in collaboration with Jean-Paul Gaudillière.

NOTES

1 Söderqvist (this volume).
2 Macquer 1777.
3 De Solla Price 1963, 70–74.
4 Söderqvist (this volume), note 33.
5 Tiedemann and Gmelin 1826, 1.
6 Berzelius 1840.
7 Berzelius 1813, 71.
8 Berzelius 1840, 60–71.
9 Roche 1828, 550–554; Poisuille 1828, 272–305 and Poiseuille 1829, 44–52.
10 Weber and Weber 1825.
11 Weber 1831, 67–71.
12 Kirsten 1986, 113–116.
13 Söderqvist (this volume).
14 Idem.
15 Olby 1974.
16 Judson 1979.
17 Cairns, Stent and Watson 1966.
18 Burian 1993, 388.
19 Rheinberger 1993, 443.
20 Watson 1968, 24–72.
21 Creager and Gaudillière 1996.
22 Newsholme and Start 1973.

BIBLIOGRAPHY

Berzelius, J.J., *Lehrbuch der Chemie* (trans. F. Wöhler), vol. 9 (Dresden, 1840).

Berzelius, J.J., "Mémoire sur la composition des fluids animaux," *Annales de Chimie* vol. 88, (1813).

Burian, Richard M., "Technique, task definition, and the transition from genetics to molecular genetics", *Journal of the History of Biology*, vol. 26 (1993), 387–407.

Creager, A. and J-P. Gaudillière, "Meanings in search of experiments and *vice-versa*: the invention of allosteric regulation in Paris and Berkeley", *Historical Studies in the Physical and Biological Sciences*, vol. 27 (1996), 1–71.

Price, Derek J. de Solla, *Little Science, Big Science* (New York: Columbia University Press, 1963).

Judson, Horace F., *The Eighth Day of Creation: The Makers of the Revolution in Biology* (New York: Simon and Schuster, 1979).

Kirsten, Christa (ed.), *Dokumente einer Freundschaft: Briefwechsel zwischen Hermann von Helmholtz und Emil du Bois-Reymond, 1846–1894*, (Berlin: Akademie Verlag, 1986).

Macquer, Pierre Joseph, *Elements of the Theory and Practice of Chemistry* (Edinburgh, 1777).

Newsholme, E.A. and C. Start, *Regulation in Metabolism* (Chichester: John Wiley & Sons, 1973).

Olby, Robert, *The Path to the Double Helix* (Seattle: University of Washington Press, 1974).

Rheinberger, Hans-Jörg, "Experiment and orientation: early systems of in vitro protein synthesis", *Journal of the History of Biology* vol. 26 (1993), 443–471.

Roche, L. Ch., "Recherches sur la force du coeur aortique; par M. Poiseuille, docteur en médecine; Dissertation inaugurale", *Archives Générales de Médecine*, vol. 18 (1828), 550–554.

Poiseuille, J-L-M. "Recherches sur la force du coeur aortique", *Journal de Physiologie*, vol. 8 (1828), 272–305.

Poiseuille, J-L-M., "Recherches sur l'action des artères dans la circulation artérielle", *Journal de Physiologie*, vol. 9 (1829), 44–52.

Stent, Gunther S., James D. Watson, and John Cairns (eds.), *Phage and the Origins of Molecular Biology* (Cold Spring Harbor, 1966).

Tiedemann, Friedrich and Leopold Gmelin, *Die Verdauung nach Versuchen* (Heidelberg, 1826).

Watson, James D., *The Double Helix* (New York: Atheneum, 1968).

Weber, Ernst Heinrich, *Friedrich Hildebrandt's Handbuch der Anatomie des Menschen*, 4th ed. (Braunschweig, 1831).

Weber, Ernst Heinrich and Wilhelm Weber, *Wellenlehre auf Experimente gegründet* (Leipzig, 1825).

CHAPTER 11

Recent Science: Late-Modern and Post-Modern

Paul Forman

INTRODUCTION

We are in the midst of a radical departure in cultural discourse, an ever more widespread disbelief in, indeed rejection of, that ever-receding 'horizon' of true knowledge, true art, true utopia which legitimated most of our cultural and social practice over the past two or three centuries. Instances of this abandonment of the modernist ambition of uniqueness, uniformity, unification, universality confront us everywhere. In every sphere of our social and cultural existences we are obliged to find our way without a vanguard, without any consensus as to which is the 'forward' direction. The demise of socialism as political and moral ideal became patent in the past ten years, though in retrospect we recognize the turn to have occurred nearly three decades ago. So also can we now recognize the *weltanschaulich* affinity of this turn with the "anything goes" aesthetic that has overwhelmed the supercilious moralism that was modernism in painting and architecture, and likewise its affinity with the epistemic relativism and latitudinarianism that has in these same three decades overturned the modern philosophic enterprise.

Why then, with such evidence at every hand, do so many knowledgable observers — and among them many of the most critical — remain sceptical of the reality and durability of postmodernity?[1] Perhaps because they see what Zygmunt Bauman points out so forcefully: that this is all intellectual 'superstructure', to cling momentarily to this terminological spar from the Marxist wrack. Beneath it, at the base, in the substructure of social and economic institutions, there is no radical reorientation — indeed, little change at all in direction. Excepting but few and still marginal practices such as recycling of waste, we operate in our daily lives under conditions of aggravated modernity. "Modern practice continues — now, however, devoid of the objective that once triggered it off"[2].

This chapter is an attempt to get a hold of postmodernity. In this effort I have gotten the most insight from the writings of Bauman, including, in particular, the insight that did postmodernity admit of essences — what it essentially refuses — that list would be headed by ambivalence, ambiguity, and irony.

> "What the inherently polysemic and controversial idea of postmodernity most often refers to (even if only tacitly) is first and foremost an acceptance of ineradicable plurality of the world. [...] By the same token, postmodernity means a resolute emancipation from the characteristically modern urge to overcome ambivalence and promote the monosemic clarity of the sameness. [...] Postmodernity is modernity that has admitted the non-feasibility of its original project. Postmodernity is modernity reconciled to its own impossibility — and determined, for better or worse, to live with it".[3]

I have spoken of 'postmodernity', but avoided the term 'postmodernism'. 'Postmodernity' is a relatively neutral term, for it is 'merely' the label of a constellation of putative actualities defining a historical reality. 'Postmodernism' is that too, but *qua* '-ism' postmodernism is also an ideology — an ideology that, Bauman points out, has arisen under conditions of postmodernity.[4] As ideology, postmodernism claims to transcend its own historical conditions, creating new standards by which to judge all preceding periods and orientations. Here the historian, like the sociologist, is obliged to maintain a degree of reserve, to take hold of postmodernism while yet holding it at arm's length, to construe postmodernism as a functional ideology responsive to the conditions of postmodernity.

Bauman sees postmodernism as an ideology of — heretofore mandarin — intellectuals, an ideology that reflects our increasing impotence as political-cultural 'legislators', increasingly unnecessary and irrelevant both to those who hold power and to those who produce our mass culture. In Anthony Giddens' powerful image, we intellectuals can no longer imagine ourselves guiding — but are now simply *riding* — a juggernaut.[5] Thus, Bauman points out,

> "the contemporary reorientation of cultural discourse can be best understood as a reflection on the changing experience of intellectuals, as they seek to reestablish their social function on a new ground in a world ill-fit for their traditional role".[6]

This interpretation is surely right, yet it remains too largely in the realm of the intellectuals' vaunting ambitions, refers too little to the world of practice — practice of the intellectuals qua *Kulturtraeger* and *Kulturerzeuger*. I hope therefore to make a small contribution to the sociology of postmodernism, by deriving this ideology of contemporary intellectuals from three leading characteristics of late-modern knowledge production, characteristics that are, in aggravated form, carried over into postmodernity: first, the ever accelerating overproduction of all cultural goods, including knowledge as perhaps the fastest-growing; second, the proliferation of an instrumentally defined and oriented science, inseparable and increasingly indistinguishable from its technological correlatives (literally, techno-science); and third, the fact, and more particularly the increasingly general acceptance of the fact, that all knowledge production, and most especially technoscience, is bound to particular interests.[7]

OVERPRODUCTION OF KNOWLEDGE AS *KULTURTRAEGERKRISE*

Modernity as intellectual orientation is conceptually coherent with, and historically inseparable from, modernity as an economic orientation. As Didier Nordon has observed,

> "society assumes that all material progress is good; the pure scientists assume that all progress in knowledge is good. Material progress is an accumulation of objects; progress in knowledge is an accumulation of objects of knowledge. In both cases, the actors produce in order to produce".[8]

What is described here is the characteristically modernist performatist compulsion, in which the activity — as measured by its output — itself becomes the purpose of the activity. "The modern world of instrumental rationality" is, as Bauman puts it, "the world where deeds are lived as means to ends and justify themselves by the ends which they serve as means".[9] Likewise technology as an end in itself is cognate and correlate in modernity with production as an end in itself. Bauman, following Jacques Ellul, points out that the capacity to do is not created in order to achieve predefined ends so much as the ends are determined by the capacity to do: 'have car, will travel'.[10]

From the point of view of a consumer society — i.e., a society composed of *passive* consumers of goods and services — there is nothing to complain of in this state of affairs. Overproduction under the conditions of our 'postindustrial' economy leads no longer to a glut of a fixed menu of 'goods', but rather to an endless augmentation of choices available, and an unceasing stimulation of the consumer (through advertising, mass media, etc.) to extend the range of his/her desires to cover those augmented choices.

If, however, we regard 'society' as comprising not passive consumers but discriminating actors, the situation appears essentially different. Then, as Herbert Simon recognized on the threshold of our 'information age':

> "In a world where information is relatively scarce [...] information is almost always a positive good. In a world where attention is a major scarce resource, information may

> [...] turn our attention from what is important to what is unimportant. We cannot afford to attend to information simply because it is there. I am not aware that there has been any systematic development of a theory of information and communication that treats attention rather than information as the scarce resource".[11]

Though presumably unacceptable as such to Simon, postmodernism is that theory. For if we give *our* attention to those special sub-societies of active cultivators of 'higher' cultural goods — i.e., those engaged, as *Kulturtraeger* and *Kulturerzeuger*, in the perpetuation and augmentation of a 'canon' of whatever sort, be it artistic, literary, scientific, or technical — then, as Randall Collins narrates:

> "What we see around ourselves in recent decades has been an enormous expansion of cultural production. There are over one million publications annually in the natural sciences, over 100,000 in the social sciences and comparable amounts in the humanities. To perceive the world as a text is not too inaccurate a description, perhaps not of the world itself, but of the life position of intellectuals: we are almost literally buried in papers. As the raw size of intellectual production goes up, the reward to the average individual goes down, at least in the realm of pure intellectual rewards of being recognized for one's ideas and of seeing their impact upon others. The pessimism and self-doubt of the intellectual community under these circumstances is not surprising".[12]

For Collins, "the content of the 'postmodern' message is to deny objective truth", and this he sees "as an ideology of cultural producers in a highly pyramided market structure, where nothing in sight seems to touch solid earth". Or, as Frank Ankersmit said, varying the metaphor slightly in motivating his essay on "Historiography and postmodernism": "My point of departure in this article is the present-day overproduction in our discipline [...] the situation in which historiography itself impedes our view of the past".[13] But metaphors apart, consider the obvious *practical* consequences of this overwhelming overproduction of higher cultural 'goods': in a situation in which the producers of culture are also the bearers of that cultural tradition, an obligation falls upon those producers to critically consider, and discriminatingly assimilate, the cultural 'goods on offer'. Under these expectations, overproduction produces a crisis of impotence.

It is, then, not that the *Kulturtraeger*, overwhelmed by this flood of paper, are metaphorically unable 'to touch solid earth', or to get a 'view of the past', but rather that they, in the face of this insuperable 'literature', must still somehow fashion a practicable mode of cultural production and reproduction. If then, we understand postmodernism not primarily as a denial of 'objective truth', but as a denial of a unique and universal canon, as primarily an affirmation of pluralism and of a plurality of standards of value, then, yes, we can very well understand postmodernism as an ideological response to the condition of postmodernity, a response that makes a virtue of that multiplication and dispersion of the producers of knowledge and products of their activity — of that superabundance which would otherwise, under modernist axioms, have to be experienced just as desperately as Collins depicts it. And if that denial of the canon is sufficiently radical, then "in a postmodern culture the writer" — I here quote a writer of a book on *Composition as a Human Science* — "needs to read what she needs in order to think, to make sense, not in order to know what is fashionable. There can be no canon of theory, any more than of literature".[14]

I quote, but I do not fail to note that there are serious difficulties with this stance, difficulties already in relation to literature, but still more serious difficulties in relation to 'the literature' of any scholarly or scientific discipline. Yet if we turn our attention to actual practice of scholarship and research in the postmodern present, we find that the modernist axioms which we historians of science have implicitly sought to fulfill in accounting for science as the 'doing' of well-disciplined disciplines have been put aside to a remarkable extent. Lawrence Dowler, associate librarian at Harvard College, starting likewise from the insuperable tide of monographic publications — he has 850,000 volumes published annually, and increasing at 2.5% per annum — concludes from his study of its use that, "The particular inquiry, rather than the academic discipline, is now the primary engine of research".[15]

LATE-MODERN/POST-MODERN KNOWLEDGE AS INSTRUMENTAL

More and more "the particular inquiry" is an 'instrumental' inquiry. It is instrumental in two senses, linking the practical and the ideological, that seem especially important to postmodernity: a scientific practice dominated by instruments, and a scientific attitude of instrumental rationality. The two together, are, in effect, the theme of Bruno Latour's *Laboratory Life* and *Science in Action: How to Follow Scientists and Engineers Through Society*. Latour's account of instruments and inscriptions finds impressive confirmation in the impressions gained in America by a visiting Russian biologist, Leonid Margolis:

> "Young scientists start to think that science consists of putting the results produced by one machine into another, and then into the next one, and of arranging thus obtained beautiful pictures and graphs into a publication".[16]

If the young scientist comes to think "technique is exciting",[17] it is because her seniors as well as her experience tell her that "science and engineering are becoming more alike, in large measure because of instrumentation", quoting Lewis Branscomb, sometime head of the National Bureau of Standards, sometime head of IBM's research, here a candidate for the presidency of the American Association for the Advancement of Science.[18]

Twenty years earlier, in the early 1960s, Derek Price, deploying his ingenious bibliometric analysis, had sought to differentiate science, as oriented toward 'the literature,' from technology, as papyrophobic. But by the early 1980s this demarcation criterion was failing. Extending Price's analysis to patents, Francis Narin suggested that:

> "There is more going on today than just a diminution of the time lag between scientific discovery and its use in technology: we suggest that high-tech patents and the related scientific papers are so tightly linked as to be almost indistinguishable, and that this is concrete evidence that high-technology and science itself are almost indistinguishable".[19]

And in reviewing recently the further studies on this question, Narin found that the average number of references in a U.S. patent to the scientific literature "has been increasing steadily from 1975".[20]

There is, of course, a technologizing of science through the deployment of instrumentation that has been essential to its operations and to its successes since the early 17th century. Indeed, Martin Harwit has argued that nearly all astronomical discoveries since the Renaissance are direct results of the introduction of new technical means of observation.[21] It is not necessary, however, to contest this secular feature of modern science in order to contend that late modernity saw a great intensification of this technologizing of the sciences, and that the enormous expansion of science since World War II has come about overwhelmingly through the creation of technoscientific tools.[22] Scientific specialties of astounding practical competence, and even entire scientific fields, have been brought into existence by an instrument or a technology. Indeed, as David DeVorkin has convincingly shown, the rocket was less the tool than the creator of 'space science',[23] which science has never been defined conceptually, but only instrumentally, namely — in the words of one of its ablest promoters, Homer Newell — as "those scientific investigations made possible or significantly aided by rockets, satellites, and space probes".[24]

In the course of these postwar decades quantitative change became qualitative transformation: the technical means ceased to be merely means and became ends as well.[25] Already in the early 1950s Otto Loewi, a physiologist of the older generation, could deplore the "general tendency of our time to worship methods and gadgets". And, Loewi continued, "This has gone so far that sometimes one has the impression that in contrast with former times, when one searched for methods in order to solve a problem, frequently nowadays workers look for problems with which they can exploit some special technique".[26]

More recently sociologist David Edge has made this "drive toward excellence by purely *technical* criteria" the burden of his study of instrumental innovation in infrared astronomy. One of his subjects began by avowing that "my own particular interests are in star formation", but came eventually to allow that:

> "I've found [...] that we always get good reviews when we do something interesting technologically. If we decide to just do science for a couple of years without developing new instrumentation, our reviews start getting worse. And I think they just like to see new technology development".[27]

The vastness and potency of technologically oriented science — and scientifically oriented technology — are of course well known. What is not so well appreciated, however, is how differently this fact is now coming to be appraised in — and in consequence of — postmodernity. It used to be that pure science, and abstract theory, had an unquestioned place of honor within the scientific estate. An applied physicist of such exceptional capacity and originality as W.W. Hansen, highly esteemed as the creator of concepts essential to the production and technical exploitation of microwaves — who died in his prime shortly after World War II — was by no means exceptional in that "pure and actually rather abstract research was held by him in almost exaggerated admiration; with typical modesty and humor he sometimes referred to his 'platonic love for pure research' ".[28]

If the American public had not shared this relative ranking of theoretical and practical science in the decades before the Second World War, they certainly did in those following. As President Kennedy said in 1963 when addressing the U.S. National Academy of Sciences at its centennial celebration:

> "If I were to name a single thing which points up the difference this century had made in the American attitude toward science, it would certainly be the wholehearted understanding today of the importance of pure science. We realize now that progress in technology depends on progress in theory; that the most abstract investigations can lead to the most concrete results [...]"[29]

Such, of course is what the Academy members wanted to believe and to hear. But the point is that the first President who was prepared to tell them that was also the last President who would be willing to tell them that: just at this moment the tide was beginning to turn.

The engineers rebelled and forced the establishment of a National Academy of Engineering. The proliferation of technoscientific microcultures, each owing its existence more to common tools than to common concepts (to say nothing of a common world view), gradually undermined the veneration enjoyed by the pure, general, and abstract theorists within the scientific estate. Just as Bauman's 'cultural legislators', the rough equivalent in the humanistic fields of science's abstract theorists, have become more and more irrelevant to the process of knowledge production, so now an analogous displacement is occurring also in the natural sciences. The postmodern mode of knowledge production — through its instrumentalism, and the priority thus assigned to specific procedures for coping with the complexity of the real world, rather than transcending that complexity through abstraction — has pushed the abstract theorists more and more out of the *scientific* limelight. However much the general public, still addicted to transcendence, continues to wonder at those who mentally "touch God",[30] the postmodern transvaluation of values (including the devaluation of disciplines) has now largely deprived those soaring theorists of the role of disciplinary culture hero.

That said, it is important to emphasize that cultural logic is never cultural inevitability, that the abstract theorists could very well have retained their standing had that been the wish of their scientific fellows — as it has remained, largely, that of the public at large. Thus here again we can identify the late-modern/post-modern transition with a cultural 'decision', taken inside as well as outside the technoscientific world, to cease venerating the pure and abstract.[31] An aggressive expression of the new — postmodern — stance comes, appropriately, from Nicolas Metropolis, a mainstay for four decades of the Los Alamos weapons-science culture:

> "[T]he fact is that quite some time ago the tables were turned between theory and applications in the physical sciences. Since World War II the discoveries that have changed the world were not made so much in lofty halls of theoretical physics as in the less-noticed labs of engineering and experimental physics. The roles of pure and applied science have been reversed; they are no longer what they were in the golden age of physics, in the age of Einstein, Schroedinger, Fermi [...]"[32]

Metropolis goes on to place in wartime Los Alamos the mythic act of recognition of the superiority of instruments over thought. By Metropolis' account, this seminal insight originated with Richard Feynman, a brilliant younger member of Hans Bethe's Theory Division:

> "Feynman soon came to realize that reliable [mechanical desk-top calculating] machines in perfect working order were far more useful than much of what passed for theoretical work in physics. [...] We spent hours fixing the small wheels until they were in perfect order. Bethe, visibly concerned when he learned that we had taken time off from our physics research to do these repairs, finally saw that having the desk calculators in good working order was as essential to the Manhattan Project as the fundamental physics".[33]

Needless to say, the advent of postmodernity was required to liberate from Los Alamos this bit of parochial culture-constituting mythology and give it global significance.

With Metropolis we are, in tone at least, already encountering the instrumental as a comprehensive ideology — an ideology that includes instrumental rationality. In postmodern science, as in war (we now think differently about love), all is fair: 'the ends justify the means'. Anathematized by liberals in modernity, this maxim is the implicit credo of postmodernity. Nowhere was the modernist rejection of this maxim clearer than in the conventional view of scientific knowledge, continually reaffirmed from the middle of the seventeenth to the middle of the twentieth century, namely that all the distinctive and desirable qualities of scientific knowledge were already immanent in the method by which it was attained, i.e., that the means sanctified the end, rather than the end justifying the means.

The switch in attitude, to an ends-justify-the-means instrumental rationality, is not without precedents, particularly among physicists, over the past hundred years.[34] But it is only recently that the conventional view has been widely and generally abandoned across the spectrum of scientific fields. These altered attitudes have an important bearing upon the much mooted question of 'scientific misconduct', the discussions of which have so often missed the mark through assumption of the conventional view that the validity of the knowledge produced follows from and depends upon the validity of the methods employed. But from a rigorously instrumentalist point of view the phrase 'scientific misconduct' can have no definite meaning — or rather the only definite meaning of which it is susceptible is 'lack of success'. Thus in the Imanishi-Kari/Baltimore case there was a clear tendency to exonerate the researchers if the published results — even if invented — prove reproducible.[35] Investigating this "murky borderland", Ullica Segerstraale found that physicists, especially, espouse such an ends-justify-the-means attitude toward scientific practice, and noted its chilling effect on whistle-blowing.[36] All this is indeed postmodern, but not specifically postmodern. Rather, it is the late-modern ethos of production as an end-in-itself, now aggravated by lifting the restraint of the modernist internal morality of means.[37]

'FOUNDATIONS' OF 'INCOMMENSURABILITY'

The ends-justify-the-means instrumentalism that has come to dominate the cultural practice of our late-modern/postmodern knowledge society has also had substantial impact upon our theorizing about the nature of scientific knowledge. The most commonly encountered periodization of the philosophy of science places in the early 1960s the beginnings of a turn away from the logical positivist view of science as a unitary, hierarchic, conceptual structure that grows by comprehending within itself an ever wider range of natural phenomena and/or adding additional layers to its hierarchy. This turn away from a disinterested, unitary, positivism in favor of a view of scientific knowledge as discovered/constructed in the pursuit of a *weltanschaulich* preconception, or in the elaboration of a preconceived program of knowledge-production, has been ascribed to Thomas Kuhn's *Structure of Scientific Revolutions* more often and more widely than to any other work. And in the philosophic literature concerned with Kuhn's theory of science, one concept, one issue, dominates discussion, viz., the putative 'incommensurability' of alternative, competitive, successive scientific theories. Kuhn himself said of 'incommensurability' in his presidential address to the Philosophy of Science Association,[38] "No other aspect of *Structure* has concerned me so deeply in the thirty years since the book was written".[39] While Kuhn, unlike most philosophers engaged in this discussion, did appreciate that the *Zeitgeist* of those three decades had a significant role in this reorientation of the philosophy of science, Kuhn, too failed to see how very largely the logical-metaphysical problem of 'incommensurability' is really just a vain reflection in philosophy of the ends-justify-the-means instrumentalism of contemporary cultural practice.[40]

THE 'BINDING' OF KNOWLEDGE[41]

"Possession of property is exclusive; possession of knowledge is not exclusive: for the knowledge which one man has may also be the possession of another [...]". This claim of non-exclusivity of knowledge, which, like the loaves and fishes, is undiminished through being shared, was already long familiar in 1886 when John Wesley Powell presented it to the U.S. Congress.[42] In the following century, however, circumstances of knowledge production and distribution inconsistent with this claim, incipient at that time, continued to multiply — until in the past decade this claim has come to seem quaint and naive in the extreme. In affirming a conception of knowledge as bound and interested, postmodernity stands in striking opposition to both the classical and the modern conception of science as a liberal pursuit, i.e., the free activity of unfettered minds, the results of which, freely published, will conduce to the freedom of all mankind.[43] Indeed, concerned as we are with notions of knowledge, this is the most distinctive criterion demarcating postmodern from modern science: postmodernity begins where the production of bound and interested knowledge is unequivocally accepted.

Proceeding in parallel since the late nineteenth century were two, only seemingly distinct, aspects of this secular change. Firstly the binding of scientific knowledge *production* to capital-possessing, self-perpetuating institutions — universities, primarily — and the binding of the *produced* knowledge to other capital-possessing, self-perpetuating institutions: government agencies, industrial firms, and more recently a wider range of corporate entities. With some few exceptions, the former process — the binding of knowledge production to institutional bases — was effectively complete quite early in this century. Although this circumstance was already emphasized by Max Weber in *Wissenschaft als Beruf*,[44] a view of knowledge as production — and consequently as importantly dependent upon its material and institutional supports — was at variance with the reigning conception of science as a liberal pursuit, pertaining only to the man and not at all to his circumstances, and consequently was strongly resisted.[45] Thus many scientific journals continued to denote their contributors' institutional affiliations in disjoint or inspecific ways well into the middle of the twentieth century.[46]

Although those journal editors and contributors alike would have insisted that there is, in any case, a great difference between an individual knowledge producer being dependent upon an institutional base for the performance of research — an undisputed fact ignored so far as possible — and the 'binding' of the produced knowledge to a particular institution or interest — a circumstance deplored wherever it could not be disputed — in the decades before and after World War II administrators of American universities showed much apprehension of the connections between these two forms of 'binding'. Through the four decades of the Cold War the binding of knowledge to a national interest and its restriction to a national territory became the leading characteristic of a very large sector — depending on definition, even the largest sector — of the R&D industry in the United States,[47] and *a fortiori* in the USSR. Although those nations' scientists — especially their exact and physical scientists — were deeply involved in classified research and embargoed technology, these features of knowledge production, lacking as they did legitimacy in the scientists' modernist ideology, were bracketed out, compartmentalized, denied any acknowledged place in their picture of how science is done, in which picture the ideal was largely made to stand in for the real.[48] Recently — very recently — this has begun to change as the modernist self-conception of a searcher after free and disinterested knowledge is exchanged for a postmodern acceptance of the legitimacy of proprietary, interested knowledges.

"At issue", *now*, "is how academic science, primarily state funded, participates in a political economy that celebrates the market". Moreover, Shiela Slaughter continues, in the United States:

> "As the presidents of elite research institutions moved segments of their universities closer to the market, they became effectively indistinguishable from chief executive officers of large corporations. They were concerned with maximizing the returns of any resources they could get for those segments of their institutions that were tied to the private sector".[49]

In this they have not only been oriented by the general shift in the political culture of the nation, but more specifically authorized and directed by new Federal and State legislation and by new policies framed by their boards of trustees. Taking the University of Arizona as a case study, Slaughter and Gary Rhoades found that between 1969 and 1989:

> "Policies and statutes moved from an ideology that defined the public interest as best served by shielding public entities from involvement in the market, to one that saw the public interest as best served by public organizations' involvement in commercial activities. Claims to the ownership and rewards of intellectual property shifted dramatically in that time, from faculty owning their products and time to complete ownership by the institution".[50]

This reorientation toward the market — which is, as such, a repudiation of the notion of 'free' knowledge — together with the increasing orientation toward the particular problem, works powerfully to dissolve the scientist's attachment to his discipline, indeed to dissolve the disciplines themselves and their disciplinary authority.[51] This is most clear perhaps in the growing acceptance of, even institutionalization of, secrecy. In late-modernity a sharp distinction was maintained between secret 'classified' research (performed principally in the interest of 'national security') and 'unclassified' ('basic', 'pure', and hence 'free') research. A principal purpose of this distinction was to enable the scientists (and philosophers of science) to treat as exceptional all that transpired in that vast realm of classified research, while maintaining the authority of the scientific disciplines in the sphere of unclassified research. In postmodernity not only does this distinction fail, but so also does the effort to maintain it. Chu, for example, in submitting his first papers on high-temperature superconductivity, falsified crucial data in order that no one else be able to begin their research at the most advanced point which he had attained.[52] In x-ray crystallography, meanwhile, it has become accepted and formalized that the atomic coordinates will not be made available to other researchers even upon publication of the alleged crystal structure, but that the 'discoverer' is *entitled* to hold them secret for one year in the U.S. and longer in Europe.[53]

In sum, the long-term historical process of binding of knowledge production to the interests of powerful institutions, which under the conditions of modernity operated under significant ideological constraints, has now, under those of postmodernity, advanced unhindered and extended in every possible direction.

POWER AND POLICY CREATE A CONVENIENT REALITY

The binding of knowledge *production* to particular institutions with their own particular interests — formerly primarily institutions with a 'military' purpose, but now increasingly market-oriented institutions — carries with it an implication, indeed an expectation, that the *produced* knowledge will 'shape reality' in ways convenient to those powerful sponsoring institutions. That shaping of reality may be through technical devices extending their physical power and/or commercial returns, or it may 'merely' be a definition of what the world contains and what is to be judged good about the world — advertising, in the widest sense — which definition advances the sponsoring institution's perceived interests. " 'In the system I work in' " — Robert Bellah and collaborators are quoting a welfare supervisor — " 'our motto could be, 'If you don't have to report it, it didn't happen'. Appearances and regulations are all that count!' "[54] What is especially to be emphasized is that as 'truth' recedes below the postmodern conceptual horizon, while instrumentalist programs of knowledge production spring up at every hand, the scientific disciplines are losing their authority as arbiters not merely of scientific conduct, but also of scientific knowledge, of the 'shapes of reality' that are possible and impossible.

We have in 'cold fusion' a striking example of this circumstance. Certainly the declining authority of the scientific disciplines is reflected in the 'discoverers' covering themselves with a cloak of secrecy — 'justified' by their intent to seek patents — and in their direct appeals to state and national legislatures for 'authorization' of their research. But the most significant loss of authority of the scientific disciplines lies, rather, in their incapability of quashing this heresy. However much the physics discipline believes cold fusion to have been disproved, however much it seeks to ostracize and ridicule cold fusion research and researchers, it is simply unable to eradicate this research program. Science journalist Gary Taubes, placing himself emphatically in the side of 'good science', wrote a thick book in which all who saw something in their experiments on cold fusion are treated as fools.[55] But at its end he has to admit that "What cold fusion had proven, nonetheless, was that the nonexistence of a phenomenon is by no means a fatal impediment to continued research. As long as financial support could be found, the research would continue".[56]

Here we enter, of course, on the most sensitive territory of postmodern doctrine, namely the linkage of power and truth. More particularly, at issue here is the reversal of the modernist reading of Bacon's "knowledge is power" — taking 'is' as an implication, rather than an equivalence — to a postmodernist 'power is knowledge', i.e., power includes the capability to create knowledge 'in its own image'.[57] That circumstance to which the postmodern reading of 'knowledge is power' points is by no means new; indeed it is far older than modernity itself, for the employment of the available knowledge producing capabilities to create a convenient reality is *in effect* the practice of all premodern, traditional societies and nearly all organized religions. In this, consequently, the critics of postmodernity who deplore it as a throw-back to pre-modernity — for example, Christopher Norris[58] — have a point, indeed they can very well point to Galileo's inquisitors and Dostoyevsky's Grand Inquisitor. They are, however, more wrong than right, for in postmodernity this process of creating a convenient reality goes forward without that unambivalent confidence in the correctness of one's own beliefs, that conviction of the absolute validity of the ends which are to justify the means, so characteristic of the pre-modern.

'INCOMMENSURABILITY' OF COMPETING PROGRAMS FOR KNOWLEDGE PRODUCTION

In *The Truth about Postmodernism*, Norris refers, impatiently, to "different (incommensurable) language-games", making it plain that such talk as Jean-Francois Lyotard's in *The Postmodern Condition* is commonplace in the postmodern parlance that Norris finds alarmingly widespread and imprudent.[59] Lyotard is indebted to Wittgenstein, of course. And a further source of the notion of incommensurability is Paul Feyerabend's writings in the sixties and seventies, where the concept appears in conjunction with an attack on modernist methodologic strictures and an advocacy of an 'anything goes' instrumentalism that is remarkably far ahead of its still not quite postmodern time.[60] Nonetheless, the quantity of formal philosophical publication on the question whether such a thing as 'incommensurablility' is possible or impossible, hence existent or nonexistent, points unquestionably to Kuhn's *Structure of Scientific Revolutions*.

My concern is not, of course, with the logical or ontological possibility or impossibility of 'incommensurability', but with the popularity that the concept has attained, which is to say, the pertinence that it is widely perceived to have for contemporary experience just now, in the period and under the conditions of postmodernity. I hope, moreover, that the reader will see in the foregoing pages several perspectives on postmodern knowledge production and distribution from which 'incommensurability' would appear advantageous to both *Kulturerzeuger* and *Kulturtraeger*. To review: Overproduction of knowledge, and insuperability of the literature embodying it, must inevitably make attractive what is at once a strategy and an excuse for ignoring the greatest part of it. That this vast expansion of the knowledge-production industry is chiefly based upon the proliferation of instruments (i.e. tools) and of sub-disciplines whose very existence originates with and depends upon said tool, itself suggests a sort of tool-constituted 'incommensurability' with conceptual, practical, and sociological dimensions.[61]

Such a tool-constituted 'incommensurability' is further reinforced, when this instrumentalism of knowledge production is carried as instrumental rationality up to the ideological level. Then, as Herbert Marcuse saw so clearly: "One does not 'believe' the statement of an operational concept but it justifies itself in action — in getting the job done, in selling and buying, *in refusal to listen to others* [...]"[62] And what indeed is this "refusal to listen to others", which I have underscored, but the operational definition of 'incommensurability'? Finally, standing in closest connection with the "refusal to listen to others" — indeed as Habermas would say, essentially a manifestation of it — is the disregard of disciplinary demands while harkening to the call of the market, or, otherwise put, the binding of knowledge production to institutions pursuing particular interests and using their power as institutions to create a convenient reality.

Thus the attractiveness of the concept of 'incommensurability' under the prevailing postmodernist adaptation to the aggravated conditions of late-modern knowledge production is that it legitimates those conditions and the adaptation to them: 'incommensurability' is then no longer a part of the problem of contemporary life but part of the resolution of that problem. The question arises of course, how far can this process go? Obviously much, much further than was ever conceivable on the basis of modernist notions of knowledge.[63] Yet when we look at painting today we have reason to wonder (and worry): here is a cultural endeavor in which, a generation ago, practitioners and critics were unified by a moralistically sanctioned consensus as to the common goal of their efforts, and consequently paid the closest attention to each other's work. And today, in the words of Larry Rivers, one of those painters who 'came afterward', "At this point no live artist is doing anything that will have the slightest influence on my work, nor will mine influence them, at least not my contemporaries".[64]

POSTMODERNITY, COMMUNITY, AND RESPONSIBILITY: RISK, TRUST, AND THE CRAVING FOR COMMUNITY

Beneath Rivers' *braggadocio* one can sense his regret, a common regret that underlies our contemporary search for 'community'. Our encounter with this word and theme at every turn, in every field and style of discourse, is the consequence of a diverse array of social and cultural problems and transformations. Postmodernity, as a failure of an accepted vision of progress, a failure of consensus regarding what is forward and what backward, what is up and what down, in social, political, and cultural development, amounts not merely to an explicit relativization of all judgements of value in these realms, but also to an implicit undermining of the traditional bases for the coherence and stability of modern culture, society, and polity. Under these circumstances, we are inclined to regard every 'community' — i.e., every congeries of distinctive purposes, practices, and markers that succeeds in giving form and meaning to a collective life of any considerable number of human beings — as more than legitimate, as a positive value. To quote Bauman once again:

> "No wonder that postmodernity, the age of contingency *fuer sich*, of self-conscious contingency, is also the age of community: of the lust for community, search for community, invention of community, imagining community. [...] Community — ethnic, religious, political or otherwise — is thought of as the uncanny mixture of difference and company, as uniqueness that is not paid for with loneliness, as contingency with roots, as freedom with certainty; its image, its allurement are as incongruous as that world of universal ambivalence from which — one hopes — it would provide a shelter".[65]

To acknowledge such yearnings is not, of course, to fail to recognize that most of the innumerable deployments of 'community' in contemporary discourse are modish, opportunistic, and disingenuous (as are also, we will shortly see, the deployments of 'responsibility'). Indeed, to a very large extent, the shibboleth of 'community' (and 'responsibility') serves the conservative purpose of defanging criticism of the economic and political order. By downplaying differences of wealth and power — of interest, generally — it enables the underlying structures for the production and maintainance of such 'goods' to continue out of sight and undisturbed. In this regard the call for community is an important contribution to postmodernity, regarded as an era in which "modern practice continues — now, however, devoid of the objective that once triggered it off".[66]

But if in recent decades in all industrial, democratic societies John Q. Public has become increasingly indifferent to disparities of wealth and power, he has also become increasingly alert to, and perturbed by, every putative threat to his own physical well-being, and especially to such 'risks' arising out of the local, national, and global production and distribution of wealth and power. Various sociologists have fixed upon risk as the typical preoccupation of contemporary society.[67] Among them is Ulrich Beck,[68] whose *Risikogesellschaft* (1986) is — or was a few years ago — "already one of the most influential European works of social analysis in the late twentieth century".[69] "The argument", wrote Beck, "is that, while in classical industrial society the 'logic' of wealth production dominates the 'logic' of risk production, in the risk society this relationship is reversed".[70] To explain without teutonic dialectic why this is so, Christopher Lash and Brian Wynne[71] — underscored by Bauman[72] — point out that modernity, especially its late phases, has created "social dependency upon institutions and actors who may well be — and arguably are increasingly — alien, obscure and inaccessible".

So construed, the preoccupation with risk derives from a lack of trust: under the conditions of late-modern and postmodern life, to an ever increasing extent we *must* trust experts and institutions, while the grounds for trust are continually eroding — specifically, we recognize ever more clearly the increasingly interested character and self-serving intents of the expert individuals and institutions on whom we depend to produce, purvey, and employ the knowledges indispensible to survival in postmodernity.[73] Otherwise stated, the decline in trust is the consequence of the general recognition that the acts of experts and institutions — indeed, those of all actors in contemporary society — are guided by a diminishing sense of *responsibility* toward those whose welfare depends upon their expert performance. It is through those ever more needful and ever less available desiderata, trust and responsibility — responsibility being the precondition for trust — that we can understand the craving for community in postmodernity. 'Community' is precisely that social relation characterized by responsibility and, consequently, trust.[74]

RESPONSIBILITY AS LEADING, AND MOST GENERAL, NORMATIVE CATEGORY

As Arie Rip has shown, 'responsibility' is a term with a complex history and a complex network of meanings.[75] For the immediate purpose, however, it suffices to distinguish two broad uses of the term. The first, with no genuine connection to the concept of community, refers to the accountability of an agent, official, or administrator for the efficient performance of assumed or assigned tasks. The second, properly communitarian, derives from the notion of responsiveness, and suggests, vaguely, attentiveness to the needs of others — of some wider or narrower class of 'others'. The former meaning predominated in modernity; the latter meaning is coming to predominate in postmodernity, as can be seen in the communitarian gloss commonly given today to the modernist, bureaucratic concept of responsibility. Thus, for example, the head of the Smithsonian Institution writes:

> To warrant continued success in attracting private resources, the Smithsonian bears a huge responsibility to the donor community. We are carefully identifying those needs that are best met through private resources and will use contemporary, yet appropriate, techniques to secure such funds.[76]

Indeed, some part of the great impact of 'responsibility' as a demand of the day results from this left-right, one-two 'punch', conflating its authoritarian with its anti-authoritarian meaning.

As we advance into postmodernity, 'responsibility' is coming to play an enormously greater role than it had in modernity — to play *both* the role of *leading* normative category and the role of *most general* normative category. "If the key word for Lyotard in 1979 was 'legitimation,' perhaps the crucial word now is 'responsibility' ".[77] And so it appears in the thematic text introducing the exhibition "Science in American Life", which opened in the Smithsonian's National Museum of American History in the spring of 1994: "The challenge of the 21st century is to make responsible choices about science and technology". Similarly, Mastercard's public service message on U.S. National Public Radio in the summer of 1995 is "use your credit card responsibly". And every reader of today's newspapers knows that "personal responsibility is the catchword" of contemporary American politics.[78]

Carol Gilligan, bolstered by feminism's affirmation of the specifically female, was one of the earliest to put responsibility forward as leading normative category. Her observation that for *women* "the moral problem arises from conflicting responsibilities rather than from competing rights",[79] was not in itself new. What was new was her elevation of the values of community, commitment, and responsibility, i.e., a transvaluation of values placing a positive sign before the feminine apprehension of the moral problem, rather than the negative sign that previously, under masculine modernist axioms, had been assigned to the feminine failure to 'understand' the primacy of individual rights. Increasingly, now, rights are being subordinated to responsibilities. A "Survey of recent work on citizenship theory" opens by noting that "there has been an explosion of interest in the concept of citizenship among political theorists".[80] After three paragraphs reviewing the changing temper of democratic polities since the late 1970s, the authors find "it is not surprising, then, that there should be increasing calls for a 'theory of citizenship' that focuses on the identity and conduct of individual citizens, including their responsibilities, loyalties, and roles", i.e., communitarian values, with no mention of rights. "Rights", we are told a few paragraphs further on, characterized "the postwar orthodoxy" but are now passé.

Such examples attest to responsibility's new role as postmodernity's leading normative category. No less significant is responsibility's new role as postmodernity's *most general* normative category, i.e., that category commonly appealed to in the widest range of ethical decisions. Indeed, in contemporary theory *and* in contemporary practice 'responsible' is commonly taken as coextensive with 'moral', as equivalent to 'ethical'. Bauman,[81] for example, leans heavily upon Emmanuel Levinas, particularly Levinas's thesis that ethics "does not supplement a preceding existential base; the very node of the subjective is knotted in ethics understood as responsibility".

Almost a decade ago in contributions to Paul Durbin's volume on *Technology and responsibility*, Carl Mitcham pointed perceptively — and with a touch of pique perhaps?:

> "To the central role the word ['responsibility'] will play in contemporary life, where "responsibility" has become a touchstone — if not cliche — in discussions of moral issues in art, politics, economics, business, religion, science, and technology. [...] The truth is that responsibility has become a general normative category [...]"[82]

Perhaps the best, the most authoritative, exemplification of this fact are the two volumes by Robert Bellah and collaborators diagnosing, and prescribing for, the cultural disease of contemporary America. *Habits of the Heart* (1985) and *The Good Society* (1991) are, to a great extent, a concerted attack on modernism — for its irresponsible individualism — and a plea for community. In the latter book we meet:

> "Marian Metzger [who] considers herself to be a *responsible* person — *responsible* in her personal life, *responsible* to the company she worked for, and *responsible* for improving the way she related to others [...] she [...] came up against the limits of her capacity for *responsible* action. [...] What she lacked was a way to think *responsibly* about the institutional forms that had brought about her quandry [...]"[83]

Though the explicit emphases are mine, responsibility is implicitly the authors' most general, as well as leading, normative category.[84]

MODERN SCIENCE AS FLIGHT FROM RESPONSIBILITY

Recognizing the central and overriding importance atttributed to responsibility not in America alone, but in all Western democratic societies in postmodernity, we must expect that these obligatory values will find some reflection in the attitudes of scientists toward themselves and their knowledge-producing activities. However, in order to appreciate how substantial a reorientation is involved in a scientist's affirmation of responsibility, it is first necessary to recognize that the traditional stance of the modern scientist, and the premodern as well, was one of radical irresponsibility.

By ancient and honorable tradition the ivory tower has been a place of retreat from individual moral responsibility, and only very rarely a support for its exercise. "I have been startled at how reluctant academics seem to be to treat ethical issues", said Clark Kerr looking back over a long career as professor of public administration and then Chancellor of the Berkeley campus, and President of the University of California.

> "As a young teacher at Berkeley [in the late 1930s], I was asked by a distinguished professor at the University of Chicago, who had a Quakerly concern about the academic study of ethics, to bring together influential members of the Berkeley teaching staff, particularly scientists, to talk with him about it. They were polite to him but not to me afterwards. They made it clear that this was not a subject which could hold any interest for scientists or scholars of any sort and that I should have known this; [...] that ethics was just a matter of personal taste and anything goes in matters of taste — with one extremely important exception: a commitment to scientific truth in the academic world".[85]

This commitment to a transcendent scientific truth overriding all moral considerations — and, in particular, overriding every responsibility — is, as Rip and Wim Smit point out, a prime example of Max Weber's "absolute ethic", the "ethic of ultimate ends".[86] In contrast with an "ethic of responsibility", which considers commitments and consequences in weighing incompatible interests and goals, an "absolute ethic", Weber wrote, "just does not *ask* for 'consequences'".[87] Such is the ethos of pure science: "I think there is little more important to the individual scientist than that there should be opportunities for him to find and to record where his mind will take him, irrespective of consequences", said Percy Bridgman,[88] surely one of this century's most reflective scientists, just at the time of Clark Kerr's early, painful experience at Berkeley.

This self-conception of truth seeker/finder as elevated above all consideration of consequences was not elaborated by the modern scientist unaided. It has been effected and sustained by the culture of modernity. One can see this today in the remnants of the Western concept of genius to be found in any middlebrow organ, but very little any longer in highbrow journals. This romantic reworking of the renaissance concept of immortal achievement,[89] includes the notion that an individual capable of producing transcendent cultural goods is, *ipso facto*, placed beyond and above the reach of moral judgements.[90] With such broad cultural encouragement to flee responsibility, it is hardly surprising that "artists, writers and generally men and women whose genius forced them to keep the world at a certain distance"[91] — scientists among them — have taken the opportunity and run.

But the institutionalization of science in Western industrial societies in the last two centuries created two further determinants of, and supports for, an ideology of irresponsibility among scientists, academic and otherwise. The first of these is the modernist, disciplinary value of production as an end in itself, to which historian of mathematics Herbert Mehrtens and sociologist of mathematics Sal Restivo have each drawn attention. As Mehrtens, especially, has emphasized, not only is modern, discipline-directed science "institutionalized irresponsibility", but, more particularly, the overriding priority given to productivity by scientific disciplines had as inevitable consequence their refusal to allow consideration of the means by which high levels of productivity are maintained, whether those means be internal to the discipline or relate rather to the terms on which social support for its knowledge-producing activities is obtained.[92]

The third main determinant/support of a scientific ideology of radical irresponsibility became effective only after the Second World War had placed the scientists' claims to social support upon the basis of 'national security'. 'National security' was not merely capable of, but insisted upon, maintaining an enormously larger scientific establishment than any previously known (and maintaining it at a much higher standard of scientific living).[93] Here now it became imperative for the scientists to fashion a self-image that allowed them to close their eyes to the real basis for the generous social support of their knowledge-producing activities and to maintain the illusion of personal and disciplinary autonomy.[94] That new self-image, originating with the American physicists and then spreading to other disciplines, projected 'fun' as the predominant feature and leading attraction of 'doing science'.[95] Its shibboleth, 'physics is fun', was meant to trigger the fantasy of perfect autonomy: any eudemonic activity is an end in itself, and as such wholly autonomous — but also wholly irresponsible. Moreover, putting 'fun' forward trumps any question of the end for which the scientist's knowledge-making is sustained, and thus supports from this side as well a stance of radical irresponsibility.[96]

IN POSTMODERNITY SCIENTISTS EMBRACE RESPONSIBILITY

Though it would require much more space to develop the case, it is important here to state that the flight of the modern scientist from responsibility was integral with modernity generally[97] — modernity understood as "a gigantic exercise in abolishing individual responsibility".[98] The scientists were permitted their stance of irresponsible purity only because such a stance was compatible with the transcendence-oriented political, aesthetic, and cognitive ideologies of modernity, and, moreover, compatible with modernity's practice of constructing ever more elaborate — ostensibly more efficient — bureaucratic social machinery in which the individual is an expert but inconscient cog. Thus each of the three determinants of modern scientists' stance pointed out above — adherence to an "absolute ethic" of truth; overriding priority of discipline-directed productivity; and refusal to be fully cognizant of the grounds for society's support — is but an aspect of those more general modernist ideologies and practices.

What, then, if these supports for scientific irresponsibility should go, separately and collectively, by the board? But just that is the case in postmodernity, as we have seen in earlier sections of this chapter: the notion of a unique, universal, transcendent truth is now incredible; the scientific disciplines have lost much of their legitimacy, and the hierarchical ranking of them in scales of abstract-practical or pure-applied has lost nearly all its authority; with the failure of truth and the end of the primacy of disciplinary demands, a 'space' is opened for moral judgements, while, simultaneously, the need for an unambiguously clean conscience, which seemed so urgent in modernity, now appears hopelessly unrealistic.

Where previously the modernist recognition of the impossibility of any *purely rational* grounding of ethical norms — conjoined with the equally modernist refusal to accept the positive, pluralist consequences of that recognition — had long embargoed ethical discourse among philosophers and scientists, social and natural, the past two decades have been "a period that might be called 'the Great

Expansion' [...] in ethics".[99] This 'great expansion' has become a great explosion in consequence of various measures by the U.S. government mandating studies and instruction in bio-medical ethics.[100] And if, on top of all this, the common culture in which scientists too are immersed adopts 'responsibility' as its leading and most general normative category, could scientists maintain — would scientists wish to maintain — their posture of irresponsibility? The answer is surely 'no', and today one finds on every hand evidence of an ideological reorientation.

A point to be noted, with respect to physicists especially, as indicative of a much greater susceptibility to pressures from the cultural environment, is the self-inculpation which they now display in facing the recent decline in financial support for 'curiousity driven' research and in employment opportunities generally. Where even just a few years ago the American physicist saw any decrease in social support as manifestation of the 'others' failure to appreciate him at his true worth, it is today commonplace for physicists to accuse themselves of various faults, particularly being arrogant, and to hold themselves in large measure responsible for their current difficulties.[101] Further it is important as a precondition, or at least concomitant, of the acceptance of moral considerations in directing their knowledge production, that these physicists now also blame themselves for holding too tightly to their disciplinary orientation. "We physicists have become exceedingly conservative in our choices of research topics", the directors of the Princeton Materials Institute and of the Institute for Theoretical Physics at the University of California, Santa Barbara, wrote recently in *Physics Today*:

> "Large numbers of investigators tend to concentrate their efforts in a few well-established areas, while other topics — often the most interesting, interdisciplinary or otherwise unorthodox ones — remain relatively untouched [...] the term 'strategic' should have an entirely positive meaning for us; there is no reason for it to have become a catchword symbolizing retreat from 'pure' or 'curiousity driven' research. Acting 'strategically' should mean simply that more of us are working on projects that are interesting not just to ourselves but also to others, particularly areas outside our own specialties. In short, it should mean maximizing our impact on the world around us".[102]

Noteworthy about this admonition is not so much the allegations of conservativism — physicists have in fact become far less, not far more, conservative in their choices of research topics — but rather the anti-disciplinary standard of progressivism that is here being applied. And this is perhaps truer still in physics paedagogy. Steven Strogatz, well known for his innovations at M.I.T., recently published a textbook based upon his course: *Nonlinear Dynamics and Chaos: with Applications to Physics, Biology, Chemistry, and Engineering*. The reviewer in *Physics Today* praised the work and noted that "the details associated with applications to lasers, pendula, fireflies, rabbits and sheep and foxes, superconductors, chemical reactions, love affairs, insect outbreaks and the coding of secret messages with chaos are self-contained".[103]

A yet more recent article in *Physics Today* provides striking evidence that the categories 'moral' and 'responsible' have indeed advanced to premiere positions in the rhetoric of this loosened discipline. Irwin Goodwin, the magazine's senior editor and a savvy analyst of physics' governmental relations, authored an account of a conference in honor of Hans Bethe's sixty years at Cornell University.[104] Titled "A tribute to a titan of modern physics", it is subtitled, "Physicists and friends celebrating Hans Bethe's scientific ingenuity and moral influence [...]" and it begins, "Even at the age of 88, Hans Albrecht Bethe is one of the world's most resourceful and responsible physicists". The prominence of 'moral' and 'responsible' is the more significant in that, to judge by Goodwin's own report, these themes appeared only in a film about Bethe screened on that occasion and not at all in the *viva voce* presentations.

Another recent, striking example of the invocation of responsibility is a lecture, delivered in the spring of 1995 at the Livermore, California, 'weapons' laboratory, by physicist Neal Lane, one year after assuming the directorship of the National Science Foundation.[105] 'Responsibility' is laced through Lane's text, appearing eight times, once every four or five hundred words. ('Values' — usually in the morally more neutral combination 'values and goals' — appeared fourteen times.) Lane's uses of 'responsibility,' although highly inspecific, retain still remnants of the word's traditional meanings. But the breadth and dominance that 'responsibility' has attained as normative category in the wider culture — where it is now commonly equated to 'moral' — is also being reproduced in the culture of science. Prompted by the rash of 'scientific misconduct' cases, and by the governmental concern about them, the U.S. National Academy of Sciences instituted a committee to say what good scientific conduct is and how it can be instilled.[106] The committee chose *Responsible Science* as title of its report, and used the word repeatedly throughout. Although the report very carefully defines 'scientific misconduct', the term 'responsible' remains undefined, perhaps just because it is employed as equivalent to everything good.[107]

It should not, however, be supposed that because the word is, as used, empty of all specific content, its use is insignificant. On the contrary, such use shows that all ethical questions are now being approached 'from the side of responsibility', so to speak. The implicit meanings of the word remain powerfully at work. This becomes clearer when we take note of the frequent appeal these days to reasonably specific notions of responsibility even without use of the word. A revealing manifesto in this regard is Howard Georgi's chapter on "Effective quantum field theories" in the imposing collection, *The New Physics*.[108] Even though his argumentation is largely traditional, his posture is 'responsible' — though the word is not used. Thus in arguing against pursuing grand unification theories, Georgi says, "their only connection with reality is through cosmology. Cosmology is fun, but [...]". Previously there would have been no 'buts'; 'fun' would have been trumps.

CONCLUSION

The foregoing exposition cannot be summarized in detail in any brief conclusion. Its principal burden is, however, that even as, in postmodernity, knowledge production becomes ever more closely integrated with the pursuit of 'special interests', and fragmented for this purpose, the present acceptance of this state of affairs as real and inevitable has also opened up the possibility — which had remained closed in modernity — for moral considerations to play a role in science. We have, as Sam Schweber has said, given up our former "belief that rigid boundaries existed between the moral and the physical domains".[109]

> "The scientific enterprise is now largely involved in the creation of novelty — in the design of objects that never existed before and in the creation of conceptual frameworks to understand the complexity and novelty that can emerge from the *known* foundations and ontologies. And precisely because we create those objects and representations we must assume moral responsibility for them".[110]

As important as these conclusions are for science, they are still more important for the historiography of science. In so saying I have in mind, of course, the obvious fact that any significant change in science is *ipso facto* also significant for us as its historians. Of far greater importance, however, are the implications of these conclusions relative to the distinction between 'us' as historians and 'them' as scientists, a distinction based fundamentally — however little we recognized it — upon the reality *in modernity* of that boundary between the physical and the moral.[111] In postmodernity, however, — to quote the words of the Chairman of the Board of the Nobel Foundation from his opening address at the 1989 awards ceremony — "[t]he steadily increasing demand for responsibility on the part alike of the researchers and the humanists is now bridging the gap between the two cultures".[112]

ACKNOWLEDGEMENTS

This chapter is an elaboration of my oral presentation at the International Workshop on the Historiography of Contemporary Science, Technology, and Medicine in Göteborg, September 1994, and employs some material that appeared in the interim confirming my argument. For helpful criticism I thank Katherine Livingston, Herbert Mehrtens, Alan Morton, and Lorenza Sebesta.

NOTES

1 Such a one was Ernest Gellner (1995), who saw only deplorable cultural manifestations, and clung to the notion of cultural and ethical truth

— Western European civilization being the closest approximation thereto — grounded in the indubitable fact of scientific truth. Similarly those banded together in the U.S. in the National Association of Scholars (Heller 1994, Browne 1995).

2 Bauman 1991, 98, 272.

3 Bauman 1991, 98.

4 Bauman 1987 and 1992a.

5 Giddins 1990.

6 Bauman 1992a, 24.

7 Only belatedly, through John Ziman's (1996) highly pertinent paper, have I become aware of the inquiry of Michael Gibbons, *et al.* (1994) into the nature and consequences of "postacademic science". Their characterization of this mode of knowledge production centers on the concept "transdisciplinarity", and includes, *inter alia*, the three key characteristics stressed in the present chapter. They fail, however, to see the interdependence of these characteristics, and they fail — perhaps through politesse — to acknowledge that "postacademic science" is not merely supplementing, but is displacing and transforming the modernist mode of knowledge production they call "academic science".

8 Nordon, *Les mathématiques pure n'existent pas!* (1981), quoted by Restivo 1990, 128–129. Cf. Kline 1980, 334, and Mehrtens 1994, 330, 339, who observes that "the leading norm [of science] is not simply knowledge, but discipline-specific productivity," and that "The self-definition and memory of mathematics force each mathematician to take his *proper* business more seriously than any other, thereby being optimally productive". These issues are a continual theme of Mehrtens' (1990) differentiation between the modern and the counter-modern in turn of the twentieth century mathematics: "Mir scheint, daß der Primat der Produktion so recht erst mit der Moderne zum Ausdruck kommt". (1990, 423).

9 Bauman 1992b, 5.

10 Bauman 1993, 187–190.

11 Simon 1978, 13.

12 Collins 1992, 92–94. Collins refers his numbers to Solla Price 1986. Cf. Gibbons *et al.* 1994, 94, 103.

13 Ankersmit 1989, 137.

14 As quoted by Schilb 1991, 182.

15 Dowler 1993. It may be objected that thirty years ago Derek J. de S. Price (1986, 56–57, 63–66, 74–76) had already recognized this problem of overproduction of scientific literature as an aspect of the exponential growth of science through three centuries of modernity, and had, moreover, 'solved' this problem by discovering science's ongoing, continual, process of growth through fragmentation into "invisible colleges". But even as we acknowledge, once again, Price's remarkable insight and prescience, we

should recognize that Price as modernist refused to feature the inevitable disciplinary disunity.

16 Margolis 1992.

17 Gornick 1983, 27–28.

18 Lewis Branscomb, Presidential candidate's statement enclosed with AAAS election ballot, 1985.

19 Narin and Noma 1985, 371.

20 Narin and Olivastro 1992, 243; Narin 1994.

21 Harwit 1981.

22 General arguments to this effect have been made by *inter alia* Forman 1987, Galison and Hevly 1992, Pestre 1992, Pickering 1989 and 1995a, and Schweber 1989. Studies entering into the details of this technologizing of late-modern science have been made by *inter alia* Baird 1993, Forman 1995a and 1996, Kay 1988, and Stine 1992.

23 DeVorkin 1996 and 1992.

24 Quoted by DeVorkin 1996, 253.

25 This has been Feasable only because the sciences are not bound as are the humanities to an ideology that requires each and every *Kulturtraeger* to carry — not the whole, but — a significant part of the whole cultural goods of the humanistic discipline. Unable to renounce this featured self-conception — because having nothing like the instrumental role of the sciences to substitute for it — the humanists have been forced to redefine "a significant part" in such a way as to cut the whole down to a manageable size. Hence post-modernism.

26 Quoted by Blume 1992, 90.

27 Edge 1992, 130, 154.

28 Bloch 1952, 125.

29 Greenberg 1967, 254.

30 Forman 1997.

31 Revealing expositions of the factors undermining the traditional hierarchy of subfields of physics: Schweber 1993b, and Cao and Schweber 1993.

32 Metropolis 1992, 120.

33 Metropolis 1992, 126-127.

34 Heilbron 1982. Schweber (1986) cites American theoretical physicist E.U. Condon as stating, in 1937, "All is fair in love and war and, I might add, in Theoretical Physics", and as referring to John Dewey in justification of his instrumentalism.

35 Travis 1993, Stone 1995, Kevles 1996.

36 Segerstraale 1990, 15.

37 Forman 1991.

38 Kuhn 1991, 3.

39 When, in *the* book on Kuhn's book, Hoyningen-Huene (1993, 207) comes to "Incommensurability", he provides some 140 references to the literature, far,

far more than he gives at any other point or for any other Kuhnian concept. And Kuhn himself, in the foreword he contributed to H-H's book, added that "Since the publication of structure my most per- sistent philosophical preoccupation has been the underpinnings of incommensurability". Biagioli (1990 and 1993, Ch. 4) has proposed an "anthropological" — i.e., trans-historical, social/logical — account of the indispensibility of incommensurability (hence, by the same rationalist reasoning that Kuhn employs, the reality of same) "derived from an homology I perceive between Kuhn's concept of paradigm and Darwin's notion of species.... According to this Darwinian metaphor, incommensurability would be *necessarily* related to the conceptual speciation of a new paradigm". Kuhn (1991, 7) found Biagioli's article "splendid", and continued to say that "With much reluctance I have increasingly come to feel that this process of specialization, with its consequent limitation on communication and community, is inescapable, a consequence of first principles". The reluctance is inseparable from the need for "first principles".

40 "But others were present too", wrote Kuhn (1991, 3), "Paul Feyerabend, and Russ Hansen, in particular, as well as Mary Hesse, Michael Polanyi, Stephen Toulmin, and a few more besides. Whatever a *Zeitgeist* is, we provided a striking illustration of its role in intellectual affairs". What that Zeitgeist might be is suggested by Steve Fuller (1992): "Being there with Thomas Kuhn: a parable for postmodern times".

41 I have borrowed the term 'binding' from John A. Remington (1988), whose perspicacity regarding the transformation under way in the production of scientific knowledge I have only recently come to recognize.

42 Dupree 1994 and 1986, 227.

43 Steven Shapin (1991 and 1994) has emphasized the extremely close connection between the early-modern English conception of the gentleman as personally free — unbeholden — and the creation there in the seventeenth century of the notion of a trustworthy scientific report.

44 Weber 1919.

45 This fiction that knowledge production is independent of its institutional locus was a specifically modern perspective, to be distinguished from postmodernity on the one side and the renaissance on the other side. In both the renaissance (Forman 1973, 155; Biagioli 1993) and in postmodernity the sponsorship of the produced knowledge is one of its principal defining characteristics.

46 Gillmor (1986) has published data showing a saltation just after World War II in the fraction of papers in ionospheric research bearing indications of the authors' institutional affiliations, a jump from just a few percent to some ninety percent. However, this dramatic rise, which supports so well the point here made, is unfortunately largely an artifact, due to inconsistencies in the source from which the data were drawn.

47 Forman 1987 and Leslie 1992.

48 Forman 1991, note 39; Forman 1996.
49 Slaughter 1993, 279, 296. Good work on this important question has been published by Eckert and Osietzki 1989, Krimsky *et al.* 1991, and, especially, Wright 1994.
50 Slaughter and Rhoades 1993, 287.
51 Gibbons *et al.* 1994, Ziman 1996.
52 Felt and Nowotny 1992.
53 Cohen 1995.
54 Bellah 1986, 126.
55 Taubes 1993.
56 Taubes 1993, 426.
57 This contention is most closely associated with Foucault, and it was taken by Lyotard as central in *The postmodern condition* (1984, 44–47). Thus Fekete (1987, 70) offers three, alternative, characterizations of "the nature of contemporary ("postmodern") culture," of which "The first is the characterization of our times that follows from the work of Michel Foucault: an age dominated by the union of knowledge and power…". Lyotard, whose historical ignorance is abysmal, and Foucault, whom I will not presume to judge, both take this postmodern reading as the one transhistorical rule of Western society since the 17th century, at least. Thus the recognition of this (transcendent) fact appears as the fruit of postmodernism, but the fact as such has no special connection with postmodernity — which, indeed, Lyotard seems rather to hope might escape precisely this condition by virtue of its postmodern insights.
58 Norris 1993.
59 Norris 1993, 23; Lyotard 1984, 53–54
60 See note 40, above.
61 Pickering 1995b, 187–189, 245–246. See also note 25, above.
62 Marcuse 1964, 103.
63 Thus on the basis of Kuhn's modernist axioms the coexistence, anyway *a la longue*, of even two, let alone many, incommensurable paradigms was logically precluded, for the existence of the scientific discipline depended on the uniqueness of the paradigm. Biagioli (1990, 207–208) — see note 39, above — approaches the question with postmodern pluralist axioms, which, however, he fails to recognize as such, and so misses the real basis of his disagreement with Kuhn.
64 As quoted from Rivers, *What did I Do*?, with A. Weinstein (New York: Harper/Collins, 1992), in a review by R. Koenig, *New York Magazine*, Nov. 2, 91–92.
65 Bauman 1991, 246–247.
66 Bauman 1991, quoted above.
67 Luhmann's book on the *Soziologie des Risikos* was prompted by "the fact that present-day society is so much concerned with risk" (1993, viii). In the index to Giddens, *Modernity and self-identity* (1991), 'risk' and 'trust' have

the largest numbers of entries, as do 'risk' and 'knowledge' in the index to Nico Stehr and Richard V. Ericson, eds, *The culture and power of knowledge* (1992). Mary Douglas, *Risk and blame* (1992, 15), asks "how to explain the new concern with risk?" Her answers show little understanding.

68 Beck 1992.

69 Lash and Wynne 1992, 1.

70 Beck 1992, 12–13.

71 Lash and Wynne 1992, 4.

72 Bauman 1993, 203.

73 For the recent, widespread fixation upon 'trust' as urgent social desideratum, see Forman (1995b).

74 That 'responsibility' trumps 'trust' appeared quite clearly during the 1992 presidential campaign, where Bush, impugning Clinton's character, pushed 'trust' as the principal issue. To which Clinton responded by charging that Bush "refused to assume a shred of responsibility" (*International Herald Tribune*, July 31, 1992, 3; as quoted by Warner 1993, 451).

75 Rip 1981.

76 Heyman 1994.

77 Jasper 1993, 2.

78 Robin Toner in *New York Times*, July 16, 1995, Section 4, 1; likewise Michael Wines, *New York Times*, Sept. 10, 1994, Section 1, 1. Such examples could be multiplied indefinitely.

79 Gilligan 1982, 19.

80 Kymlicka and Norman, 1994.

81 Bauman 1992c, 43.

82 Mitcham 1987a, 3; Mitcham 1987b, 361.

83 Bellah *et al.* 1991, 22–23.

84 Bellah *et al.* 1986, Bellah *et al.* 1991. The truth is that the solution Bellah *et al.* offer (e.g. 1991, 40–41) to this quandry — how indeed to think and act responsibly in relation to the extremely powerful and intrinsically amoral institutions of our society — comes pretty close to fascism: institutions are for them beautiful things, embodiments of the highest ideals, etc.; in their index the word 'bureaucracy' does not appear.

85 Kerr 1989, 139.

86 Arie Rip and Wim A. Smit, "Toward a responsible university: taking societal impacts of scientific research into account", (Paper presented at Seminaret i Vetenskapteori, University of Oslo, February 15, 1991).

87 Weber 1946, 120.

88 Bridgman 1938, 250.

89 Forman 1973.

90 To take a couple of examples that came randomly to hand in the *New York Times Book Review* (March 22, 1992, Section 1, 29; March 8, 1992, Section 1, 11): Andrew Sullivan, editor of the *New Republic*, reviewing a bio-

graphy of the Australian novelist Patrick White, sees this undeniably mean and wretched character as "a vessel for extraordinary artistic and spiritual transcendence"; Walter Moore, author of a biography notorious for its revelations of Erwin Schroedinger's irresponsible personal behaviors, reviewing David Cassidy's biography of Heisenberg, finds nothing unusual in the deficiencies of *this* physicist's personal character — "Scientists under all regimes tend to prefer expediency to morality" — but stresses, as his bottom line, that "Although Heisenberg's political life was far from heroic, his idealistic interpretation of physics will nonetheless endure".

91 The phrase, used by Hannah Arendt's biographer as descriptive of her, appears here as quoted by Thomas Söderqvist (1991, 151–152).

92 Mehrtens 1994. Cf. note 8, above.

93 Forman 1987.

94 Forman 1996.

95 Forman 1989.

96 This is the sense of Norbert Wiener's lament over the "degradation of the position of the scientist [...] to that of a morally irresponsible stooge in a science factory". Quite another dimension of irresponsibility in postwar science was stressed by various scientists caught up in the transition to 'big science.' To Melvin Schwartz, for example, it seemed that "in such a 'production-line' approach to scientific research '*nobody* need feel real responsibility for the accuracy of the results.'" Both quotations from Capshew and Rader, 1992, 11.

97 Forman 1997.

98 Bauman 1992a, xxii.

99 Darwall *et al.* 1992, 123.

100 Thus 1989 in National Institutes of Health required that all institutions receiving its training grants provide instruction in research ethics to apprentice researchers, and more recently the funds for deciphering the human genome included a 'set-aside' for ethical inquiries.

101 Striking evidence of this is to be found in the statements made in two "roundtable" discussions published in *Physics Today* (Coppersmith *et al.* 1994) and (Byer *et al.* 1995).

102 Eisenberger and Langer 1995.

103 Ronald F. Fox in *Physics Today*, vol. 48 (March 1995), 93–94.

104 Goodwin 1995.

105 Neal F. Lane, "The scientist in an age of unreason", lecture delivered at Lawrence Livermore National Laboratory, (May 18, 1995), typescript, 8 pp.

106 U.S. National Academy of Sciences 1992.

107 The report includes a second, supplementary, volume of commissioned background papers, none of which seeks to define responsibility. The committee, though it did not define 'responsible', did state the "fundamental

responsibility" of the "individual scientist", namely, "to ensure that their results are reproducible, and that their research is reported thoroughly enough so that results are reproducible" — an instrumentalist criterion that says nothing about the individual's conduct in research or toward others, and that, moreover, has rarely ever been fully met, and is increasingly deliberately unmet, as we have seen above. And while noting ironies, I cannot refrain from pointing out that spokesmen for the modernist resistance typically cannot themselves avoid appealing to postmodern normative categories. Thus Ernest Gellner (see note 1) says "to pretend that the scientific revolution of the seventeenth century, and its eventual application in the later stage of the industrial revolution, have not transformed the world, but are merely changes from one culture to another, is simply an *irresponsible* affectation". And Gerald Holton, who gets the last word in the New York Times's account of the May 31–June 2, 1995, "The Flight from Science and Reason" conference in New York (Browne 1995), calls American scientists to arms to defend "the *moral* authority of science". (Emphases are mine, of course.) For more postmodern-despite-themselves examples, see Forman (1995b).

108 Georgi 1989.

109 Schweber 1993a.

110 Schweber, 1993b.

111 Forman 1991.

112 Nobel Foundation 1989, 7.

BIBLIOGRAPHY

Ankersmit, Frank R., "Historiography and postmodernism", *History and Theory*, vol. 28 (1989), 137–153.

Baird, Davis, "Analytical chemistry and the 'big' scientific instrumentation revolution", *Annals of Science*, vol. 50 (1993), 267–290.

Bauman, Zygmunt, *Legislators and Interpreters, On Modernity, Post-modernity and Intellectuals* (Cambridge: Polity Press, 1987).

Bauman, Zygmunt, *Modernity and the Holocaust* (Cambridge: Polity Press, 1989).

Bauman, Zygmunt, *Modernity and Ambivalence* (Ithaca, N.Y.: Cornell University Press, 1991).

Bauman, Zygmunt, *Intimations of Postmodernity* (London: Routledge, 1992a).

Bauman, Zygmunt, "Survival as a social construct", *Theory, Culture and Society*, vol. 9 (1992b), 1–36.

Bauman, Zygmunt, *Mortality, Immortality and Other Life Strategies* (Stanford: Stanford University Press, 1992c).

Bauman, Zygmunt, *Postmodern Ethics* (Oxford: Blackwell, 1993).

Bauman, Zygmunt, *Life in Fragments: Essays in Postmodern Morality* (Cambridge Mass: Blackwell, 1995).

Beck, Ulrich, *Risk Society: Towards a New Modernity* (London: SAGE, 1992).
Bellah, Robert N. *et al. Habits of the Heart: Individualism and Commitment in American Life* (1985; New York: Harper and Row, 1986).
Bellah, Robert N., *et al. The Good Society* (1991; New York: Vintage Books, 1992).
Biagioli, Mario, "The anthropology of incommensurability", *Studies in History and Philosophy of Science*, vol. 21 (1990), 183–209.
Biagioli, Mario, *Galileo, Courtier: The Practice of Science in the Culture of Absolutism* (Chicago: University of Chicago Press, 1993).
Bloch, Felix, "W.W. Hansen", *Biographical Memoirs of the National Academy of Sciences*, vol. 27 (1952), 120–137.
Blume, Stuart, "Whatever happened to string and sealing wax?" pp. 87–101 in Robert Bud and Susan E. Cozzens (eds.), *Invisible Connections: Instruments, Institutions and Science* (Bellingham, WA: SPIE Optical Engineering Press, 1992).
Bridgman, Percy W., *The Intelligent Individual and Society* (New York: Macmillan, 1938).
Browne, Malcolm W., "Scientists deplore flight from reason", *New York Times* (June 6 1995), C1, C7.
Bud, Robert and Susan E. Cozzens, eds., *Invisible Connections: Instruments, Institutions and Science*, (Bellingham, WA: SPIE Optical Engineering Press, 1992).
Byer, Robert L. *et al.* "Roundtable: wither now our research universities?" *Physics Today*, vol. 48 (March 1995), 42–51.
Cao, Tian Yu and Silvan S. Schweber, "The conceptual foundations and the philosophical aspects of renormalization theory", *Synthese*, vol. 97 (1993), 33–108.
Capshew, James H., and Karen A. Rader, "Big Science: Price to the present", *Osiris*, vol. 7 (1992), 3–25.
Cohen, Jon, "Share and share alike isn't always the rule in science", *Science*, vol. 268 (June 23, 1995), 1715–18.
Collins, Randall, "On the sociology of intellectual stagnation: the late twentieth century in perspective", pp. 73–96 in M. Featherstone (ed.), *Cultural Theory and Cultural Change* (London and Newbury Park, CA: SAGE, 1992).
Coppersmith, Susan N. *et al.* "Physics roundtable: reinventing our future", *Physics Today*, vol. 47 (March 1994), 30–39.
Darwall, Stephen, Allan Gibbard, and Peter Railton, "Toward fin-de-siècle ethics: some trends", *Philosophical Review*, vol. 101 (1992), 115–189.
DeVorkin, David H., *Science with a Vengence: The Military Origins of the Space Sciences in the American V-2 Era* (New York: Springer, 1992).
DeVorkin, David H., "The military origins of the space sciences in the American V-2 era", pp. 233–260 in P. Forman and J.M. Sanchez-Ron (eds.), *National Military Establishments and the Advancement of Science and Technology: Studies in Twentieth Century History* (Dordrecht: Kluwer, 1996).
Douglas, Mary, *Risk and Blame: Essays in Cultural Theory* (London: Routledge, 1992).
Dowler, Lawrence, "Scholars, technology, and library resources", *Perspectives* (Newsletter of the American Historical Association), (April 1993), 16–19.
Dupree, A. Hunter, *Science in the Federal Government* (Baltimore: Johns Hopkins University Press, 1986).
Dupree, A. Hunter, "A knowledge policy for peace", *Technology in Society*, vol. 16 (1994), 289–300.

Eckert, Michael and Maria Osietzki, *Wissenschaft für Macht und Markt: Kernforschung und Mikroelektronik in der Bundesrepublik Deutschland* (Munich: Beck, 1989).

Edge, David, "Mosaic array cameras in infrared astronomy", pp. 130–167 in Robert Bud and Susan E. Cozzens (eds.), *Invisible Connections: Instruments, Institutions and Science* (Bellingham, WA: SPIE Optical Engineering Press, 1992).

Eisenberger, Peter and James S. Langer, "Opinion: a case for strategic research", *Physics Today*, vol. 48 (April 1995), 78–80.

Fekete, John (ed.), *Life after Postmodernism: Essays on Value and Culture* (New York: St Martin's Press, 1987).

Felt, Ulrike and Helga Nowotny, "Striking gold in the 1990s: The discovery of high-temperature superconductivity and its impact on the science system", *Science, Technology, and Human Values*, vol. 17 (1992), 506–531.

Forman, Paul, "Scientific internationalism...", *Isis*, vol. 64 (1973), 151–180.

Forman, Paul, "Behind quantum electronics: national security as basis for physical research in the United States, 1940–1960", *Historical Studies in the Physical Sciences*, vol. 18 (1987), 149–229.

Forman, Paul, "Social niche and self-image of the American physicist", pp. 96–104 in M. De Maria *et al.* (eds.), *The Restructuring of Physical Sciences in Europe and the United States, 1945–60* (Singapore: World Scientific, 1989).

Forman, Paul, "Independence, not transcendence, for the historian of science", *Isis*, vol. 82 (1991), 71–86.

Forman, Paul, " 'Swords into ploughshares': breaking new ground with radar hardware and technique in physical research after World War II", *Reviews of Modern Physics*, vol. 67 (1995a), 397–455.

Forman, Paul, "Truth and objectivity", *Science*, vol. 269, (1995b), 565–567, 707–710.

Forman, Paul, "Into quantum electronics: the maser as 'gadget' of Cold War America", pp. 261–326 in P. Forman and J.M. Sanchez-Ron (eds.), *National Military Establishments and the Advancement of Science and Technology: Studies in Twentieth Century History* (Dordrecht: Kluwer, 1996).

Forman, Paul, "Transcendence, or the flight from responsibility: modern science in postmodern perspective", in P. Galluzzi (ed.), Proceedings of the conference *Scienza & Potere*, Florence, December 1994 (to appear, 1997).

Fuller, Steve, "Being there with Thomas Kuhn: a parable for postmodern times", *History and Theory*, vol. 31 (1992), 241–275.

Galison, Peter and Bruce Hevly, eds., *Big Science: The Growth of Large-scale Research* (Stanford: Stanford University Press, 1992).

Gellner, Ernest, "Anything goes", *Times Literary Supplement* (June 16, 1995), 6–8.

Georgi, Howard M., "Effective quantum field theories", pp. 446–457 in P. Davies (ed.), *The New Physics* (Cambridge and New York: Cambridge University Press, 1989).

Gibbons, Michael, Camille Limoges, Helga Nowotny, Simon Schwartzman, Peter Scott, and Martin Trow, *The New Production of Knowledge: The Dynamics of Science and Research in Contemporary Societies* (London, etc.: SAGE Publications, 1994).

Giddens, Anthony, *The Consequences of Modernity* (Stanford: Stanford University Press, 1990).

Giddins, Anthony, *Modernity and Self-identity: Self and Society in the Late Modern Age* (Stanford: Stanford University Press, 1991).

Gilligan, Carol, *In a Different Voice: Psychological Theory and Women's Development* (Cambridge, MA: Harvard University Press, 1982).

Gillmor, C. Stewart, "Federal funding and knowledge growth in ionospheric physics, 1945–81", *Social Studies of Science*, vol. 16 (1986), 105–133.

Goodwin, Irwin, "Bethe fest: a tribute to a titan of modern physics", *Physics Today*, vol. 48 (June 1995), 39–41.

Gornick, Vivian, *Women in Science* (New York: Simon and Schuster, 1983).

Greenberg, Daniel, *The Politics of Pure Science* (New York: New American Library, 1967).

Harwit, Martin, *Cosmic Discovery: The Search, Scope, and Heritage of Astronomy* (New York: Basic Books, 1981).

Heilbron, John, L., "*Fin-de-siècle* physics", pp. 51–73 in C.G. Bernhard, *et al.* (eds.), *Science, Technology, and Society in the Time of Alfred Nobel* (Oxford: Pergamon, 1982).

Heller, Scott, "At conference, conservative scholars lash out at attempts to 'delegitimize science,' " *Chronicle of Higher Education* (November 23, 1994), A18.

Heyman, I. Michael. "Smithsonian perspectives", *Smithsonian* (December 1994), 12.

Hoyningen-Huene, Paul, *Reconstructing Scientific Revolutions: Thomas S. Kuhn's Philosophy of Science* (Chicago: University of Chicago Press, 1993).

Jasper, David. "Introduction: religious thought and contemporary critical theory", pp. 1–6 in D. Jasper (ed.), *Postmodernism, Literature and the Future of Theology* (New York: St. Martin's Press, 1993).

Kay, Lily E., "Laboratory technology and biological knowledge: the Tiselius electrophoresis apparatus, 1930–1945", *History and Philosophy of the Life Sciences*, vol. 10 (1988), 51–72.

Kerr, Clark, "The academic ethic and university teachers: a 'disintegrating profession'?" *Minerva*, vol. 27 (1989), 139–156.

Kevles, Daniel, J., "The assault on David Baltimore", *The New Yorker* (May 27, 1996), 94–109.

Kline, Morris, *Mathematics: The Loss of Certainty* (New York: Oxford University Press, 1980).

Krimsky, Sheldon, James G. Ennis, and Robert Weissman, "Academic-corporate ties in biotechnology: a quantitative study", *Science, Technology, and Human Values*, vol. 16 (1991), 275–287.

Kuhn, Thomas S., "The road since structure", pp. 3–13 in Arthur Fine, Micky Forbes, and Linda Wessels (eds.), *Proceedings of the 1990 Biennial Meeting of the Philosophy of Science Association* (East Lansing, MI: Philosophy of Science Association, 1991).

Kymlicka, Will and Wayne Norman, "Return of the citizen: a survey of recent work on citizenship theory", *Ethics*, vol. 104 (1994), 352–381.

Lash, Scott and Brian Wynne, "Introduction" to Ulrich Beck, *Risk Society: Towards a New Modernity* (London and Newbury Park, CA: SAGE, 1992).

Leslie, Stuart W., *The Cold War and American Science* (Cambridge, MA: MIT Press, 1992).

Luhmann, Niklas, *Risk: A Sociological Theory* (Berlin and New York: W. de Gruyter, 1993).

Lyotard, J-F., *The Postmodern Condition: A Report on Knowledge* (Minneapolis: University of Minnesota Press, 1984).

Marcuse, Herbert, *One-Dimensional Man: Studies in the Ideology of Advanced Industrial Society* (Boston: Beacon Press, 1964).

Margolis, Leonid B., "Does American science need Russian humanitarian aid?" *The New Biologist*, vol. 4 (1992), 413–417.

Mehrtens, Herbert, *Moderne-Sprache-Mathematik: eine Geschichte des Streits um die Grundlagen der Disziplin und des Subjekts formaler Systeme* (Frankfurt A.M.: Suhrkamp Verlag, 1990).

Mehrtens, Herbert, "Irresponsible purity: the political and moral structure of mathematical sciences in the National Socialist state", pp. 324–338, 411–13 in M. Renneberg and M. Walker (eds.), *Science, Technology and National Socialism* (New York: Cambridge University Press, 1994).

Metropolis, Nicholas, "The age of computing: a personal memoir", *Daedalus*, vol. 121 (1992), 119–130.

Mitcham, Carl, "Responsibility and technology: the expanding relationship", pp. 3–39 in P.T. Durbin (ed.), *Philosophy and Technology, vol. 3: Technology and Responsibility* (Dordrecht: Reidel, 1987a).

Mitcham, Carl, "Responsibility and technology: a select, annotated bibliography", pp. 361–387 in P.T. Durbin (ed.) *Philosophy and Technology, vol. 3: Technology and Responsibility* (Dordrecht: Reidel, 1987b).

Narin, Francis, "Patent bibliometrics", *Scientometrics*, vol. 30 (1994), 147–155.

Narin, Francis and E. Noma, "Is technology becoming science?" *Scientometrics*, vol. 7 (1985), 369–381.

Narin, Francis and Dominic Olivastro, "Status report: linkage between technology and science", *Research Policy*, vol. 21 (1992), 237–249.

Nobel Foundation, *The Nobel Prize* (Stockholm: Almqvist & Wicksell International, 1990).

Norris, Christopher, *The Truth about Postmodernism* (Oxford: Blackwell, 1993).

Pestre, Dominique, "Les physiciens dans les sociétés occidentales de l'après-guerre: un mutation des pratiques techniques et des comportements sociaux et culturels", *Revue d'Histoire Moderne et Contemporaine*, vol. 39 (1992), 56–72.

Pickering, Andrew, "Pragmatism in particle physics: scientific and military interests in the post-war United States", pp. 174–183 in F.A.L. James (ed.), *The Development of the Laboratory: Essays on the Place of Experiment in Industrial Civilization* (London: Macmillan, 1989).

Pickering, Andrew, "Cyborg history and the World War II regime", *Perspectives on Science*, vol. 3 (1995a), 1–48.

Pickering, Andrew, *The Mangle of Practice: Time, Agency, and Science* (Chicago: University of Chicago Press, 1995b).

Price, Derek J. de Solla, *Little Science, Big Science ... and beyond*, R.K. Merton and E. Garfield, eds. (New York: Columbia University Press, 1986).

Remington, John A., "Beyond big science in America: the binding of inquiry", *Social Studies of Science*, vol. 18 (1988), 45–72.

Restivo, Sal, "The social roots of pure mathematics", pp. 120–143 in S.E. Cozzens and T.F. Gieryn (eds.), *Theories of Science in Society* (Bloomington: Indiana University Press, 1990).

Rip, Arie, *Maatschappelijke verantwoordelijkheid van chemici*, doctoral dissertation, University of Leiden (privately printed, 1981).

Schilb, John, "Cultural studies, postmodernism, and composition", pp. 173–188 in P. Harkin and J. Schilb (eds.), *Contending with Words: Composition and Rhetoric in a Postmodern Age* (New York: The Modern Language Association of America, 1991).

Schweber, S.S., "The empiricist temper regnant: theoretical physics in the United States, 1920–1950", *Historical Studies in the Physical Sciences*, vol. 17 (1986), 55–88.

Schweber, S.S., "Some reflections on the history of particle physics in the 1950s", pp. 668–693 in L.M. Brown *et al.* (eds.), *Pions to Quarks* (Cambridge: Cambridge University Press, 1989).

Schweber, S.S., [Review of W. Lanouette, *Leo Szilard*], *Science*, vol. 261 (1993a), 1461–1462.

Schweber, S.S., "Physics, community and the crisis in physical theory", *Physics Today* vol. 46 (November 1993b), 34–40.

Segerstraale, Ullica, "The murky borderland between intuition and fraud", *International Journal of Applied Philosophy*, vol. 5 (Spring 1990), 11–20.

Shapin, Steven, "'A scholar and a gentleman': the problematic identity of the scientific practitioner in early modern England", *History of Science*, vol. 29 (1991), 279–327.

Shapin, Steven, *A Social History of Truth: Civility and Science in Seventeenth-Century England* (Chicago: University of Chicago Press, 1994).

Simon, Herbert A., "Rationality as process and as product of thought", *American Economic Review*, vol. 68, nr 2 (1978), 1–16.

Slaughter, Shiela, "Beyond basic science: research university presidents' narratives of science policy", *Science, Technology, and Human Values*, vol. 18 (1993), 278–302.

Slaughter, Shiela and Gary Rhoades, "Changes in intellectual property statutes and policies at a public university: revising the terms of professional labor", *Higher Education*, vol. 26 (1993), 287–312.

Söderqvist, Thomas, "Biography or ethnobiography or both? Embodied reflexivity and the deconstruction of knowledge-power", pp. 115–142 in F. Steier (ed.), *Research and Reflexivity* (London: SAGE, 1991).

Stehr, Nico and Richard V. Ericson (eds.), *The Culture and Power of Knowledge: Inquiries into Contemporary Societies* (Berlin and New York: W. de Gruyter, 1992).

Stine, Jeffrey K., "Scientific instrumentation as an element of U.S. science policy: National Science Foundation support of chemistry instrumentation", pp. 238–263 in Robert Bud and Susan E. Cozzens (eds.), *Invisible Connections: Instruments, Institutions and Science*, (Bellingham, WA: SPIE Optical Engineering Press, 1992).

Stone, Richard, "Baltimore defends paper at center of misconduct case", *Science*, vol. 269 (1995), 157.

Taubes, Gary, *Bad Science: The Short Life and Very Hard Times of Cold Fusion* (New York: Random House, 1993).

Travis, John, "Imanishi-Kari says her new data shows she was right", *Science*, vol. 260 (1993), 1073–1074.

U.S. National Academy of Sciences, National Academy of Engineering, and Institute of Medicine, Committee on Science, Engineering, and Public Policy, Panel on Scientific Responsibility and the Conduct of Research, *Responsible Science: Ensuring the Integrity of the Research Process*, 2 vols. (Washington, D.C.: National Academy Press, 1992).

Warner, Daniel, "An ethic of responsibility in international relations and the limits of responsibility/community", *Alternatives: Social Transformation and Humane Governance*, vol. 18 (1993), 431–452.

Weber, Max, *From Max Weber: Essays in Sociology*, H.H. Gerth and C.W. Mills, eds. (New York: Oxford University Press, 1946).

Wright, Susan, *Molecular Politics: Developing American and British Regulatory Policy for Genetic Engineering, 1972–1982* (Chicago: University of Chicago Press, 1994).

Ziman, John, "Is science losing its objectivity?" *Nature*, vol. 382 (1996), 751–54.

CHAPTER 12

Scientists as Policymakers, Advisors, and Intelligence Agents: Linking Contemporary Diplomatic History with the History of Contemporary Science

Ronald E. Doel

"No modern statesman can afford to be scientifically illiterate"
(Abba Eban, Chief Israeli delegate to the United Nations, Ambassador to the United States, circa 1960)[1]

INTRODUCTION: 'SCIENCE IN BLACK'

After 1940 — which A. Hunter Dupree has identified as the beginning of a Great Instauration for science in America — scientists entered the realm of foreign policy as never before.[2] The development of the atomic bomb has come to symbolize the union between scientists and the U.S. political establishment, but atomic energy was only one of many international issues that faced scientists and policymakers in the decades after Hiroshima. As the Cold War deepened, U.S. scientists labored to increase their presence in and influence over foreign policy, creating an Office of International Relations within the National Academy of Science, promoting a system of science attachés within the Department of State, and organizing science offices within the Central Intelligence Agency and its sister organizations; they also evaluated science as a potential agent of psychological warfare and urged new programs of scientific aid to Europe, Latin America, and Africa. These expanding roles for science and scientists profoundly shaped the growth of scientific disciplines, institutions, and the professional identities of researchers. Statesmen and government officials, who faced increased reliance on experts previously unrepresented in the coalitions that forged U.S. foreign policy, found these changes no less transforming.

Much of this history remains unwritten, however, and our perceptions of the role that science and scientists played in the Cold War are constrained as a result. This chapter addresses the historiographic and practical challenges of studying the international science after 1940, particularly its integration into the planning and execution of foreign policy. Here I shall concentrate on the U.S. experience between 1945 and 1960, although these arguments can also be applied to other national contexts and even more contemporary times. Many points of contact existed between scientists and foreign policy in this period: besides the dominant issue of atomic weapons diplomacy, scientists joined business leaders, industrialists, international cultural groups, and other elements of American society to press for favorable access to foreign markets or relief from international regulation.[3] Moreover, 'international science' has several distinct but overlapping meanings, embracing individual collaborations and participation in international scientific unions as well as deliberate efforts by governments to utilize science and scientists to reach foreign policy goals.

In this chapter, I shall concentrate on the intersection of scientists with establishments of foreign policy at the highest levels of U.S. government, particularly the efforts of scientists and public officials to employ science as a means to advance U.S. diplomacy during the early Cold War era. My aim, to put it another way, is to call attention to a potential history of 'science in black' — a history that focuses less on the visible, public institutions of science and the published achievements of their scientists than on the large, unexplored continent of interconnections, maintained in secrecy, between scientists and public officials mutually interested in adopting science to serve U.S. interests and the national security state. Such a history, by focusing on poorly studied aspects of the post-1945 period, can have enormous significance in re-evaluating how the Cold War affected democratic institutions and established centers of science, and provide new insight into ways that science influenced, and was influenced by, U.S. foreign policy.[4]

HIDDEN ASPECTS OF POSTWAR SCIENCE: THE INTERNATIONAL ARENA

In keeping with the central theme of this book: what, in historiographic terms, is new about science in foreign policy after 1940? Since the eighteenth century historians have traced increased overlap between the establishments of science and the state, resulting from the emerging vigor of nation-states and ascendant utility of practical science. To be sure, the Enlightenment concept of science as a 'Republic of Letters', beholden to no particular national identity, was slow to fade. In 1780, British forces allowed a Harvard expedition to attempt observations of a solar eclipse in Maine's Penobscot Bay despite the Revolutionary War, and in 1813 Sir Humphry Davy travelled to Paris to accept a prize from the Académie des Sciences in the midst of the Anglo-French military conflict.[5] Nevertheless, as Charles C. Gillispie has argued, modern government-funded weapons research was already taking form by 1793 when Claude-Louis Berthollet, the founder of physical chemistry, conducted secret investigations with gunpowder; successful application of this work in the *levée en masse* of 1793–94 helped establish the French Republic.[6] Throughout the nineteenth and early twentieth centuries, scientists strengthened their international ties by creating international scientific unions and award systems such as the Nobel Prize; furthermore, by the 1920s, the International Education Board, one of the largest of the Rockefeller Foundation philanthropies, had made significant investments in scientific research overseas.[7] At the same time, scientists became ever more closely associated with nationalism and foreign policy aims. World War I, the 'chemist's' war, marked an important milestone. Allied scientists excluded German colleagues from the newly formed international unions in 1919 for their famous "Appeal to the Civilized World" of 1914, supporting the Kaiser's decision to back the Austrian attack on Serbia that led to the world's first global conflict.[8]

Yet World War II marked a distinct watershed in the integration of scientists into the political establishment, in the United States as well as other scientifically advanced nations. The building and use of atomic weapons in 1945 had enormous influence on foreign policy, and implications drawn from the mushroom clouds over Hiroshima and Nagasaki, raising fears about cataclysmic destruction, have justly received considerable attention from historians of science and diplomatic historians alike.[9] Nonetheless, World War II incorporated science and scientists into foreign policy in two additional ways, both lacking clear precedent. First, it led to the creation of new, *permanent* institutions designed to assess or manipulate international science to benefit U.S. national security. By 1950 these came to include the Office of Scientific Intelligence of the Central Intelligence Agency (CIA), the Science Advisor's office within the Department of State, and the Research and Development Board (RDB) of the Department of Defense,[10] as well as myriad homes for international science planning within other government branches, such as the Psychological Strategy Board (PSB) the Operations Coordinating Board (OCB), the National Security Council (NSC), and development agencies such as the International Cooperation Administration (ICA) and the Agency for International Development (AID). Such agencies created new *professional* roles for scientists.

Second, the prestige value of scientific and technological advances became an important issue in the conduct of foreign policy, a linkage forged in the early 1950s but greatly strengthened after the launch of Sputnik in 1957. These points, of course, do not exhaust the range of novel themes requiring treatment. A full account of U.S. involvement in international science after 1945 must assess, in addition to atomic diplomacy, and military research and development commitments, how estimates of scientific advances in friendly and hostile nations influenced foreign policy decisions, how anxieties involving biological and chemical warfare affected international diplomacy, how and why leaders of the National Academy of Sciences (NAS) and the National Science Foundation (NSF) initiated science programs aimed at

Latin America, and how covert international activities concealed within the budgets of mainstream centers of U.S. science such as the National Institutes of Health affected the mandates and morale of these agencies. Also demanding attention is the integration of science and technology within the Point IV provisions of the Marshall Plan, the gift of nuclear reactors by the Atomic Energy Commission (AEC) to friendly Third World nations, the creation of new career options for scientists, the effects of covert science on peer review procedures, and the extent to which the demands of the post-war national security state transformed international practices in science, substantially as well as symbolically.

To date, however, there has been scant historical work on these issues. Science after 1940, as Arnold Thackray has recently remarked, remains "a sort of last frontier in the history of science".[11] Nevertheless, several approaches are promising. Scholars have examined the sharp rise of federal and military patronage for science in the Cold War, charting ways that funding influenced the character and missions of research universities and steered the interests of researchers towards problems of practical or potential military interest; the evolution of M.I.T. into a home for secret military-academic studies of national security, for example, cannot be understood outside the international context of the Cold War.[12] Other scholars have investigated scientists in government advisory agencies, including Gregg Herken's study of presidential science advising and Daniel J. Kevles's analysis of the Korean War's impact on American science policy.[13] New work has also appeared on institutionalized secrecy in post-war centers of civilian-military research and how loyalty investigations affected scientists in the 1950s.[14] International science has been a central focus for Elisabeth Crawford (social, economic, and political contexts of transnational science), Aant Elzinga (the multi-national politics of Antarctic research), John Krige (European physics and space consortiums), and Yakov Rabkin (U.S.-U.S.S.R. scientific exchanges).[15] Yet many important dimensions of science in foreign policy remain unexplored. Official histories of federal science agencies with important international dimensions, such as the AEC, have provided at best tantalizingly brief glimpses at the coordination of science with foreign policy.[16] The role of major philanthropic foundations — so critical for international science relations in the early twentieth century — is also poorly understood in the post-1945 era.[17]

Similarly, international science has not attracted the attention of many diplomatic historians, except in the case of atomic policy. Studies of Truman, Eisenhower, and Kennedy administrations, not unexpectedly, have focused on the application of science-based technologies to warfare and its role in bipolar Cold War defensive capabilities, including the strategic implications of NSC 68, Eisenhower's New Look policy, and the origins of nuclear overkill.[18] But a number of scholars, using various methods, have begun to explore how scientists, as members of an expert class, gained enough power to create, dominate, shape, or influence U.S. foreign relations during the early Cold War era. Using the corporatism approach of Michael J. Hogan, which assigns priority to the relationship between large business associations and related government agencies in the making of foreign economic policy,

John Gimbel assessed U.S. reparations policy on science and technology towards Germany immediately following World War II, while Richard Damms has called attention to the industrial connections of scientists who advised Eisenhower on national security policy.[19] National security policy — succinctly defined by Melvyn P. Leffler as "actions deemed imperative to protect domestic core values from external threats" — is central to John Lewis Gaddis' recent examinations of overflight surveillance, John Prados' study of the NSC decision-making, Walter Laqueur's analysis of scientific intelligence, Joseph P. Manzione's work on scientists and international scientific relations in the 1950s, and Robert A. Divine's investigations of the Nuclear Test Ban Treaty debate and the U.S. response to Sputnik; the prestige value of science in foreign policy has been treated by several scholars as well, including the cultural historian Michael Smith.[20]

Other approaches in diplomatic history also hold much promise for assessing the influence of international science. Michael Hunt's emphasis on ideology as an important component of foreign policy merits attention, particularly given the popular identification of science with liberal democratic values and the vivid cultural images that shaped American attitudes towards nuclear energy.[21] Bureaucratic politics (explored in the works of Leslie Gelb and J. Garry Clifford) may also illuminate the encouragement (and resistance) that scientists experienced in the State Department, the CIA, and related governmental agencies.[22] Yet relatively few studies in diplomatic history have focused on the role of science in these agencies. New information on scientific intelligence within the CIA has recently come to light through the publication of declassified, sanitized early histories of this agency by professional historians. But these volumes address administrative issues, providing little information about how scientific intelligence affected decision-making within the NSC, the training and quality of individuals working within scientific intelligence, how such material was analyzed, or the kinds of research efforts that CIA researchers undertook.[23]

Nevertheless, by neglecting the role of science in international affairs, historians risk overlooking critical aspects of U.S. foreign policy in the Cold War era — studying the tip of the iceberg while ignoring a wide array of interconnected agencies, private foundations, individuals, attitudes and shared beliefs that, until now, have largely remained below sea level. Several examples may help make this clear. It is increasingly apparent that scientists struggled to obtain a voice in the State Department after 1945, and did so in conscious efforts to promote foreign policies favorable to their disciplinary and professional interests. Their campaign for international scientific cooperation was neither wholly altruistic nor based on popular notions of science divorced from politics. The famous 1950 *Science and Foreign Relations* report to the State Department (written at the request of Under Secretary of State James Webb by Lloyd V. Berkner) defended international science as a way to strengthen Western economies, promote democratic values, and enhance U.S. research. This report, we now know, was accompanied by a secret supplement, still classified, detailing the value of employing civilian and diplomatic channels to obtain scientific intelligence intended to aid national security planning. Yet the integration of scientists into the political fabric became increasingly contentious during the Eisenhower Administration, and by 1959 debate rose in the NSC over international science policy regarding outer space, Antarctica, contacts with individuals in non-recognized regimes, and the two-Chinas problem.[24] Until recently, the schism between scientists and the U.S. government after the mid-1960s was thought to be rooted in opposition to the Vietnam War.[25] But these cracks appeared much sooner in foreign policy debate, an important point for scholars studying the Gaither Committee report on American nuclear defenses of 1957, the start of Nuclear Test Ban Treaty negotiations the following year, or the role that scientists played in the Vietnam conflict.

When scientists are viewed as active participants in forging foreign policy coalitions, additional, less obvious links between core national security issues and international activities remote from them sometimes become apparent, casting other Cold War activities in a new and quite different light. An important case in point is the International Geophysical Year (IGY) of 1957–58. Involving scientists from sixty-seven countries in coordinated studies of the earth sciences, the IGY seemed to contemporary writers and historians a remarkable thaw in Cold War tensions.[26] Walter McDougall's 1985 *The Heavens and the Earth* made clear that Eisenhower's interest in the IGY had much to do with using a civilian scientific satellite, a centerpiece of IGY planning, to establish a legal precedent for orbital overflight of the Soviet Union. More recently, in an important new examination of Berkner, the geophysicist and prime mover of the IGY, Allan A. Needell notes that Berkner first proposed this massive undertaking the same month *Science and Foreign Policy* reached the State Department. Needell argues that it is impossible to understand State Department and military backing for this major, much-heralded program without understanding the degree to which covert intelligence-gathering, national security objectives, and belief in the relationship of science with Western values were mutually linked for scientists and government officials.[27] This is not to argue that the IGY failed to enhance international scientific cooperation or to produce important geophysical insights, for it did; yet it is important to underscore the extent to which the IGY was characteristic of, and not an exception to, U.S. international science policy in the 1950s. The Cold War cast deep shadows across virtually all aspects of international science, more than many historians of science or diplomacy have recognized; assessing scientific programs of this kind can thus illuminate how U.S. policymakers saw international science as a vehicle to promote American values and interests in the post-war world.

Careful attention to scientific intelligence can also help illuminate how the United States evaluated the cost of possible Soviet expansion into Western Europe during the early Cold War era, a critical factor in understanding the origin of mutual security pacts such as the North Atlantic Treaty Alliance (NATO). As Gaddis and Robin Winks have recently reminded diplomatic historians, a critical 'so what' question must be addressed: did intelligence really matter to policy? "Fascination with esoteric minutiae" gathered through espionage, Gaddis notes, can make one lose sight "of what, if anything, it all meant".[28] One case where scientific intelligence *may* have mattered involves Sweden in the late 1940s. Sweden's neutrality policy led its government to reject membership in NATO, a step the United States had desired to solidify Scandinavian defenses against Soviet expansionism. After the NSC debate in December 1948 over possible challenges to Sweden's neutrality policy, the Truman administration let this issue drop from view, despite mounting East-West tensions over the Berlin blockade and the outbreak of the Korean War in 1950.[29] This situation makes the recently declassified OSI 1/49, "An Estimate of Swedish Capabilities in Science" — the first major report by the CIA's Office of Scientific Intelligence — an important find. Over sixty pages in length, addressing Swedish achievements in biological warfare, forestry, solid-state physics, metallurgy, and Pulsejet engine and axial-flow turbine technology, this assessment declared that the Soviets "would gain distinct, but relatively small, advantage from the point of view of science by overrunning Sweden at the present time". To what degree this report, regarded by Swedish historians of science as comprehensive and highly accurate, actually influenced Truman administration policy towards Scandinavia must still be determined.[30] But its recommendation to deny 20 to 40 top Swedish scientists to Soviet authorities if hostilities arose indicates that the once contentious policy that guided Project Paperclip, which brought German rocket engineers to the United States in 1946, were now integrated into national security planning.[31]

Finally, further attention to the role of international science as an element of psychological strategy is warranted. Until recently, historians have argued that Truman and Eisenhower administration officials paid little attention to scientific prestige in psychological warfare campaigns until the international implications of Sputnik's launch in October 1957 made their strategic value apparent. In part such judgements rested on earlier studies of the PSB and the OCB, which declared that high levels of bureaucratic infighting and ineffective coordination limited their influence on foreign policy.[32] But newly declassified documents from these agencies suggests a more pervasive role. Rip Bulkeley has shown that the OCB, emphasizing psychological factors, was a significant actor in developing Eisenhower Administration policy toward space flight and Antarctica by the mid-1950s. In addition, Martin J. Medhurst has recently argued that the Eisenhower-era Atoms for Peace program must be understood as a psychologically-based global strategy that permitted Eisenhower to implement his New Look doctrine, activities in which the OCB and the Atomic Energy Commission played major roles.[33] Greater attention to the role that major science-based federal agencies played in carrying out U.S. foreign policy is needed. What it meant for impoverished developing countries, with limited scientific infrastructures, to receive gifts of research reactors through Atoms for Peace, is a related problem that also merits inquiry.

This is hardly an exhaustive list. Other broad areas of high-level international science policy are just beginning to emerge from the shadows of Cold War secrecy. Bio-chemical warfare so worried Harvard President James B. Conant that in 1946 he advised a correspondent that "if B.W. is really a weapon like the Atomic Bomb", a public demonstration like the Bikini test might be necessary. Despite important new work on this issue by Barton Bernstein, James Hershberg, and Susan Wright, how bio-chemical weapons affected U.S. foreign policy during the early Cold War — particularly following 1952 Communist charges that NATO forces had employed biological and chemical agents in Korea — has received little attention from historians of science or diplomacy.[34] Scientific and technical aid programs to Western Europe (and later Latin America and Africa) have also received little attention; it would be good to know for example how ICA or AID programs influenced the environmental sciences, or public perceptions of the dangers of nuclear testing and chemical pollution.[35] Cold War diplomatic tensions also buffeted international scientific activities far from core national security areas and matters of high policy. Frank Ninkovich has demonstrated that by 1950 the State Department, adjusting to Cold War pressures, succeeded in making cultural internationalism an instrument of national policy — a dramatic reversal of its pre-war role as a promoter of private initiatives. We need to understand how these changes affected international meetings of scientists, and foreign perceptions of U.S. researchers.[36] All of these topics are fruitful areas of inquiry for historians of diplomacy no less than historians of science.

UNCOVERING THE HISTORICAL RECORD: DIFFICULTIES AND OPPORTUNITIES

One reason why historians have not yet turned to the intersection between science and foreign policy after 1940 involves source materials: available documentation is meager. In fairness, it must be noted that certain record groups, including those of the Office of International Relations within the National Academy of Science, have simply been under-utilized. Moreover, federally funded research centers such as RAND often distributed expurgated reports on international projects to maintain the professional credibility of their researchers, and several presidential science advisors did publish memoirs.[37] Yet many significant records remain in locked vaults, since they contain material still presumed vital to national security. Researchers interested in atomic weapons policy, one of the most frequently visited topics in the history of modern science, still find large numbers of documents partially or wholly classified.

The situation is worse for topics where few historians have worked, since the number of documents already released through researcher-initiated mandatory reviews or Freedom of Information Act (FOIA) requests is smaller. It appears that about 70 percent of documents relating to international science housed at the Truman and Eisenhower presidential libraries remain classified; significant portions of declassified materials were sanitized before release. Of greater concern to scholars, however, is that several relevant record groups, including virtually all State Department and CIA files on international science activities, remain within their originating agencies. Historians wishing access to these materials must determine, often without the advantage of finding aids, the identity of relevant records, then demand their release through FOIA requests. None of this proceeds quickly, a vexing problem for graduate students needing access to source materials with minimal delays. Documents that must be cleared by the NSC require, at this writing, 18 to 36 months for decisions to be made. Review times for Federal Bureau of Investigation (FBI) records, another important source of information on official attitudes towards scientists and international science, can be longer still.[38]

But this is not the only archival problem that confronts historians interested in international science during the Cold War. University archivists, guided by the prevailing interests of historians of science, have preferentially preserved the papers of science faculty noted for their contributions to disciplinary research, rather than their international science assignments. These categories certainly are not exclusive: major collections exist for George Kistiakowsky, the noted Harvard chemist who became Eisenhower's Science Advisor in 1959, and for Frank Press, the eminent earth scientist who led geophysical studies for the Nuclear Test Ban Treaty and later became Science Advisor to President Jimmy Carter.[39] Nevertheless, many scientists who addressed foreign policy issues after 1940 no longer worked at universities at all but in agencies and institutions lacking open archives; as a result, little information is available on individuals who spent much of their careers in the NSC, RDB, the CIA, the ICA, or the National Security Agency. Even when scientists involved in foreign policy work remained at university posts, and donated their papers to appropriate repositories, significant gaps in documentation often appear. Because 'black' science activities by their nature were covert or classified, files about them are often missing or significantly thinned. The extensive collection of Roger Adams, the University of Illinois chemist who served as science advisor to General Lucius Clay in postwar Germany, seems to contain few or no records classified as "confidential" or higher. While important records of Adams' service under Clay do exist, few records have survived of Adams' equally significant role as a senior advisor on foreign science to other government agencies.[40]

The same limitations characterize the papers of H.P. Robertson, the Caltech physicist and Science Advisor to SACEUR (Supreme Allied Commander, Europe), and those of Vannevar Bush, a principal architect of post-World War II military science policy and scientific intelligence gathering.[41] Certain individual's collections richly document the relationship of individual scientists to the CIA: for example, the Gerard P. Kuiper papers illuminate this leading American astronomer's activities as an advisor on Soviet astronomical and aeronautical research.[42] The Frank Press collection provides important insights into the roles played by geophysicists in negotiations for the Nuclear Test Ban Treaty.[43] Yet these seem exceptions. The difficulty of storing classified papers in academic offices thus limits the kind of archival material needed to investigate how individual scientists participated in covert and international science, and the extent of these activities within the general scientific community. Archival collections for contemporary scientists often document their professional lives less completely than those for senior colleagues who concluded their careers before 1940. The same holds true for the papers of non-scientists, including administrators and lawyers, who took on critical oversight or organizational responsibilities within Presidential review panels such as the first Hoover Commission of 1948 and the Second Hoover Commission of 1955.[44]

Moreover, historians of science face extraordinary challenges in investigating scientific work that was neither conducted openly nor funded through traditional channels. Standard tools of the trade, including citation counts, memoirs, and professional biographies, are of limited value for studying scientists active in 'black' fields of international science.[45] Official appointments are sometimes misleading. Wallace R. Brode, the nationally prominent chemist and former Ohio State professor who served as Associate Director of the National Bureau of Standards from 1947 to 1958, actually occupied this post only after October 1948. Before then, under an agreement worked out between NBS Director Edward U. Condon and Admiral Roscoe C. Hillenkoetter, director of the newly established CIA, Brode used this appointment as a blind while he organized scientific intelligence within this agency. Such practices were not uncommon in Cold War Washington.[46] In addition, historians of science have often appeared reluctant to press scientists to discuss their connections to 'black' science activities in oral history interviews, partly recognizing the reality of Cold War restrictions but also, one suspects, because historians of science are often uneasy about raising issues of evident discomfort to their subjects.[47] The ending of the Cold War has made it only slightly easier for scientists to reflect on the contradictions and professional compromises that affected international relations in science.

Nevertheless, the prospects for research in this field are good. Determined researchers will find a variety of resources already available. Papers from numerous members of Congress in office during the Cold War have been deposited, chiefly in university archives, around the country. Since these collections include records of discussions with individual scientists on international relations, and often unguarded comments about the politics and utility of including scientists in diplomatic efforts, they are a valuable (and heretofore virtually untapped) source for the role of science in foreign relations.[48] Conflicts within the international science unions, often reflecting government involvement and interference and foreign policy objectives, are often documented in the papers of union officers. Records for the Office of International Relations of the National Academy of Science are available at the Academy's Washington-based archives, which also contain related files on international science activities. Copies of many declassified government and agency records are available at the university-based National Security Archive in Washington, whose staff can provide invaluable assistance to historians investigating science-state relationships during the Cold War. The two-to-three year delay in obtaining documents at presidential libraries should not discourage historians already in secure positions, since mandatory review requests often release volumes of new information: at the Truman library, for instance, approximately 70 to 80 percent of material sought in this way is subsequently cleared for release.[49]

One of the most important sources of information on science in foreign policy during the Cold War has until recently been extremely under-utilized: oral history interviews with science attachés, State Department officials, federal scientists, administrators, and others who witnessed new science-state relationships as they emerged. The value and limitations of oral history for writing the history of science, which was recently treated at length, do not need to be repeated here.[50] Yet few oral history interviews have explored at depth interrelations between scientists and foreign policy. For instance, the extensive oral history collection at the Truman Presidential Library contains few interviews with scientists active in the Truman Administration, and other administration officials familiar with international science policy, including John R. Steelman, typically were not questioned on this subject. Oral history programs organized by historians of science have generally done little better at addressing this issue, focusing instead on disciplinary, professional, or largely national topics.[51] The occasional oral history that does address the relationship between scientists and foreign policy — for instance, Elizabeth Hodes' interview with the chemist Joseph B. Koepfli, who served as Science Advisor to Secretary of State Dean Acheson between 1950 and 1952, done as part of Caltech's ongoing oral history program — demonstrates how effective in-depth interviews on this topic can be.[52]

Despite concern that too many actors in international science policy from the early Cold War era are already deceased — there is no argument that oral history projects on this topic should have been launched earlier — much oral history work can still be accomplished, and should be vigorously pursued. Several justifications for doing oral history interviews in history of science are particularly applicable here: archival sources are relatively thin and sometimes deliberately obscure; interviews can act as guides to still-classified or voluminous record groups; and finally, interviews often uncover records, whose eventual preservation is one of the greatest benefits of oral history undertakings. Many individuals once active in this area remain alive and mentally active. Preliminary interviews completed this summer make clear the potential rewards of treating such issues as tensions over CIA operations and the attitudes of State Department officials towards science and scientific developments.[53] Oral history interviews are especially valuable also for exploring what was tacitly understood among participants and 'went without saying' in written documents. Questions particularly worth posing to scientists — in part because so little documentation is available for these topics — include their motivations for extending science policy to Latin America and sub-Sahara Africa, their impressions of covert international science activities, and their views on the value of maintaining contacts with colleagues in Communist China, the Soviet Union, Eastern Europe, and other areas under Communist domination.

Finally, the role of science in foreign policy is an inherently international subject; diplomatic historians as well as historians of science have advanced convincing arguments for treating such issues in comparative analytical frameworks to address such complex issues such as dependency, imperialism, and ideology.[54] To date, relatively little work on scientists and foreign policy has been done in European or Asian contexts after 1945, although several studies have begun to emerge. For instance, Klaus-Harro Tiemann has investigated exchanges of scientists between the Soviet Union and East Germany in militarily critical fields of earth science during the 1950s, while Nikolai Krementsov has explored how Stalin and leading Politburo members sought to employ perceived breakthroughs in cancer treatment as a political instrument to counter the U.S.'s atomic bomb monopoly in 1946 and 1947.[55] Archives in Europe and the former Soviet Union (including, for example, the Academy of Sciences of the former East German state) have yet to be mined carefully for international science in the Cold War.[56] Ironically, with much documentation on U.S. activities in foreign science during the early Cold War era still classified and unavailable, these archives may yield abundant documentation and insights into Western science policy, and provide the added bonus of Eastern perspectives on these actions.

REAPPRAISING THE HISTORY OF SCIENCE AFTER 1940

Much work must be done before cohesive new interpretations of international science in the Cold War era can be evaluated and revised. Nevertheless, it is worth framing significant questions even if resources to answer them are not yet fully at hand. In this section I will suggest several lines of investigation that merit further consideration by historians of science and diplomatic historians alike. In keeping with the aim of this book, these questions are biased towards history of science, although they cannot be adequately addressed without parallel efforts by historians of U.S. foreign policy.

How did political ideology affect the views of U.S. scientists towards international science during the Cold War? Until recently, historians who have studied the involvement of scientists in foreign policy during the early Cold War years have focused on 'outsiders', typically physicists, who sought to use the moral authority of international science to aid democratic causes. In the U.S., attention has centered on such individuals as Linus Pauling and Leo Szilard, who opposed McCarthyist attacks, restrictions on international travel, and development of the hydrogen bomb. In the Soviet context, similar and deserved attention has been devoted to the influential dissonant Andrei Sakharov, like Pauling a winner of the Nobel Prize for Peace.[57]

The importance of atomic diplomacy certainly makes this an important aspect of Cold War history. Yet we know far less about scientific 'insiders' who became involved in foreign policy through the State Department, CIA, NSC, and other institutions. Edward Teller was hardly the only conservative physicist in the United States after World War II, nor was physics the only science discipline whose members struggled to define U.S. foreign policy in the post-war years. For example, we know comparatively little about the influence of the Caltech astrophysicist Howard Percy Robertson, who served as science advisor to General Alfred M. Gruenther (Supreme Allied Commander, Europe) and moved freely among defense and foreign advisory positions in the United States. Similarly, we know little about Wallace R. Brode, the first head of scientific intelligence within the CIA (1947–48) who in his role as Science Advisor to the State Department (1958–60) attempted to limit U.S. scientific contacts with scientists from non-recognized regimes in order to preserve the clarity of U.S. foreign policy towards Formosa and Communist China.[58]

Hunt's emphasis on ideological factors in U.S. foreign policy raises an important question: how well can the decisions and career choices of these individuals be described in terms of ideology, and what do these actions suggest about the extent of political coherence and dissent within the broad community of U.S. scientists in the 1950s? At present, through the work of Spencer R. Weart, Loren R. Graham and others, we have a better understanding of the sharp political struggles between liberal and conservative scientists, and their consequences for research programs and foreign policy, for France and the Soviet Union, where fractures in Cold War politics became visible early in the Cold War.[59] There is every reason to expect that careful study of the intersections between U.S. scientists and such establishments as the State Department, the NSC, and foreign development agencies will yield a richer and more illuminating portrait of scientists in foreign relations than determined focus on nuclear issues has heretofore permitted. Needell's recent study of Project TROY, a 1950 study of psychological warfare funded by the State Department, provides such a window into the interlocking network of scientists, government policy-makers and military leaders forged through national security concerns — and has the added benefit of addressing social scientists as well, whose contributions to Cold War research have thus far received limited attention.[60] Similar studies are needed of leaders of the major philanthropic foundations during the early Cold War era, who often carried out foreign policy initiates that established agencies were unable or unwilling to do. The Ford Foundation's ties to the CIA, and the willingness of Foundation managers to fund U.S. participation in the international scientific unions after Secretary of State John Foster Dulles balked at doing so, make it a prime candidate for study.[61]

To what degree did U.S. scientists see international science as a significant tool in promoting democratic tendencies abroad? One of the most consistent arguments that U.S. scientists used to justify contacts with Soviet counterparts during the Cold War was that science was a wedge capable of strengthening advocates of democracy behind the Iron Curtain. The extent to which this belief was realized in practice still awaits examination; partially declassified NSC records provocatively hint that this policy bore fruit in Czechoslovakia in the decade following the 1948 Communist takeover.[62] But another interesting and far less studied environment where U.S. scientists sought influence during the 1950s was the Third World, particularly Latin America and Africa. The collapse of the colonial empires in the aftermath of World War II, and the Point IV provisions of the Marshall Plan, which called for "a bold new program for making the benefits of our scientific advances and industrial progress available for the improvement and growth of underdeveloped areas", gradually turned the focus of development plans from Western Europe towards the Third World.[63] By the early 1950s, leaders of the newly formed NSF were actively campaigning for opportunities to fund research in Latin America, and by the end of that decade, acting through the leadership of the President's Scientific Advisory Council (PSAC), the National Academy of Sciences established commissions to study science in Latin America as well as sub-Sahara Africa.[64] To what degree did proposed Latin American programs reflect needs of U.S. researchers designing research programs for the IGY (1957–58), and to what extent did they reflect attitudes and values consistent with the Monroe Doctrine?[65] Did scientists take the lead in advocating technical support for Africa, particularly within developmental agencies such as the International Cooperation Administration, or is their participation more a response to political opportunities?[66] Few studies have examined the extent to which these programs fulfilled the aims of U.S. foreign policy, or shaped the world-views and professional identities of scientists.[67]

How did covert funding for international science policy affect existing, overt centers for science during the Cold War? How did it influence peer review and decisions about the direction of major research programs? Although important studies have been made of classified military projects designed to aid U.S. national security, including Michael Aaron Dennis' work on the role of secrecy and its reconstitution of the boundary between civilian and military, what I have in mind are endeavors intended to influence foreign policy or to aid intelligence-gathering, rather than weapons development. These efforts did not merely shift scientific and human resources into agendas remote from those at pre-war universities, as Stuart W. Leslie argues in his important study of Cold War research at M.I.T. and Stanford,[68] but undermined the integrity of public institutions of science that placed barriers not merely between lower-level researchers and classified materials but also before chief administrators and their superiors. It has only recently come to light, for instance, through fragmentary references, that agency heads at the National Institutes of Health and NSF complained about providing official cover for covert science programs disguised within their budgets.[69]

That such compromises were made to fight the Cold War should not be surprising, but these practices, made under nominal peacetime conditions, raise many unanswered questions: were peer review processes bent or bypassed in accommodating 'black' science? What percentage of science agency awards during the height of the Cold War represented covert missions? How did these undertakings affect the funding, research missions, morale, and independence of these agencies? One way of approaching these questions may be to take use Chandra Mukerji's insight, developed in her study of the militarily strategic field of oceanography, that scientists became an "elite reserve labor force", on tap to meet pressing national emergencies.[70] We now know that a 1953 CIA assessment of Soviet Scientific and Technical Manpower, which found that the quantity and quality of Soviet research to be "approaching comparability with that of the United States", circulated among key Eisenhower Administration officials.[71] How important were such estimates in securing federal funds for major scientific undertakings such as the IGY? Understanding how foreign policy concerns affected highest level decisions about the cultivation of scientific research within Cold War America will complement existing studies of military patronage's influence on domestic disciplinary developments, including Paul Forman's now classic study of the development of quantum electrodynamics after World War II.[72]

How did the pursuit of international science in the U.S. context differ from that in other nations? World War II and the Cold War caused leaders of many Western governments, including Great Britain, West Germany, and France, to place science attachés in their foreign embassies, to develop agencies responsible for scientific intelligence, and to negotiate treaties to govern the exchange of scientists and strategic information.[73] As the historian Regis Cabral has shown in the case of the Brazilian-German nuclear agreement of 1954–55, and Aant Elzinga in his study of the 1950s debates over the sovereignty of Antarctica, international science became a political issue of new and unprecedented importance in Brasilia, Madrid, Stockholm, Buenos Aires, and Bonn as well as Washington and Moscow.[74] Too little is known about these relationships in any national context to permit meaningful transnational comparisons, but there are reasons for caution in any effort to boldly extrapolate the U.S. experience to other nations. For example, the relationship between international geophysics and scientific intelligence that Lloyd Berkner saw as a fundamental aspect of the IGY was not mirrored north of the 49th Parallel. Geophysical prospecting in the vast expanse of Canada's northern regions, not military planning, was a main concern for leading Canadian geophysicists such as Toronto's J. Tuzo Wilson; the international opportunities presented by the IGY were considered in a quite different political context in Ottawa than in Washington.[75] Examining how scientists resolved the opposing pulls of nationalism and internationalism during the Cold War, done within a nationally comparative context, should increase our understanding of how scientists saw their self-identity and social allegiances, and how particular political environments shaped their professional outlook.

What motivated scientists to take part in scientific intelligence activities during the Cold War? Another, and perhaps more rewarding way, to put this question is to ask: how temperamentally suited were scientists to take part in espionage? This is not meant to privilege scientists as particularly moral, but rather to suggest that the professional reward structures of science were (and are) deeply at odds with the requirements of covert intelligence-gathering — a potentially serious flaw in maintaining consistently high quality information and analysis. As the sociologist Robert K. Merton argued in his classic study of scientific norms, free exchange of scientific information in return for recognized priority seems a necessary foundation for the progress of science.[76] Did the Cold War represent a protracted period during which the Mertonian norms of science were violated? Archival evidence hints that it was far easier to recruit U.S. scientists to work as occasional informants or consultants for the CIA than to recruit scientists to temporary intelligence assignments in Washington, and still more difficult to find scientists willing to give up research or teaching careers for permanent positions within the CIA, particularly as U.S. involvement in the Vietnam War escalated. Yet scientific intelligence often proved a critical issue for NSC members in the 1950s, who struggled to come to terms with estimates of Soviet bio-chemical and medical research, tabulations of mushrooming East Bloc research capacities, estimates of potential losses of facilities and scientists in countries bordering the Iron Curtain in the event of Soviet invasion, and the relationship between proposed scientific projects and their international prestige value. As Sallie Pisani emphasizes in her recent study of the CIA and the Marshall Plan, those who contribute or control information available to policy makers themselves play an important role in setting policy.[77] No work adequately addresses the limitations of scientific intelligence after 1945, although it is now clear that Truman administration officials recognized weaknesses and sought more effective monitoring of foreign scientific developments.[78] It is important, therefore, to assess the quality of scientific intelligence received by the NSC. The high accuracy of the CIA's 1949 Science in Sweden reports suggests that, at least at that time, exacting standards were not impossible to obtain.

To continue along these lines: what fraction of the U.S. scientific community participated in CIA work, apart from the mandatory debriefings of scientists returning from overseas conferences? Were certain disciplinary communities more involved in covert international science than others — and if so, what social, professional, or national security factors stimulated their involvement? Is it coincidental, or significant, that several leading scientists who took part in the Office of Scientific Intelligence and related organizations were chemists by training, perhaps reflecting government concern about chemical warfare or the strength of the Soviet chemical industry? Or did this concentration reflect the professional experiences of chemists, more closely involved with industrial research activities than, for example, astrophysicists?[79] Did political outlook matter, and are there ways of characterizing participants along demographic or generational lines? Answering such questions will provide new insights into professional and social factors that affected U.S. scientific communities during the Cold War.

Undoubtedly new questions will be raised as new studies of these issues are made. But addressing international science and its various nationally-situated foreign policy contexts will yield a history of modern science that is better rooted to the context of the postwar era than historiographic approaches based on continuity with pre-1940 period have thus far permitted.[80] This approach will require historians of American science to take seriously Charles Rosenberg's reminder that "We cannot understand the modern world without an understanding of the necessary connections between the individual and his or her discipline, between the discipline and the social sources of its support, between ideas and their real impacts in a real world".[81] Scientists working within the fabric of the U.S. political establishment after 1945 can be examined as part of the expanded role of experts within foreign policy councils, or as agents of ideology, or participants in corporatist strategies, or as innovators in economic strategies to enhance Western geopolitical standing. Diplomatic historians are at home in pursuing such questions. Increased dialogue between historians of science and diplomatic historians is needed to achieve sharper insights into the role and significance of science and scientists in international affairs.

ACKNOWLEDGEMENTS

I wish to thank Marion Deshmukh, David H. DeVorkin, Katharine Doel, Carl-Henry Geschwind, Gregg Herken, Melvyn P. Leffler, Jefferson P. Marquis, Allan A. Needell, Anna K. Nelson, R. Allen Smith, Robert W. Smith, Jessica Wang and Spencer R. Weart for comments on earlier versions of this essay, as well as fellow participants at the International Workshop on the Historiography of Contemporary Science, Technology, and Medicine, held in Göteborg, Sweden, September 1994. I would also like to thank archivists at the Truman and Eisenhower presidential libraries, particularly Dennis Bilger and Dwight Strandberg, for their extremely helpful suggestions for locating relevant documents and requesting mandatory reviews. I gratefully acknowledge support for aspects of my own work presented herein from the National Endowment for the Humanities, National Science Foundation grant NSF SBR-9511867, the Smithsonian Institution, the Center for History of Physics of the American Institute of Physics, and the Herbert C. Pollock Award.

NOTES

1 Eban, quoted in James Killian, "Making Science a Vital Force in Foreign Policy", n.d. [circa 1962–63], (Box 2, Ludwig Audrieth papers, University of Illinois archives).
2 Dupree 1972, 443.
3 For example, the geologist and administrator Everette deGolyer, who discovered the fecundity of the Saudi Arabian oil fields while on a U.S. foreign assignment in 1943, played an important role in alerting State Department officials to the value of these deposits; see Yergin 1992, 392–93. Scientists also worked with diplomatic officials on occasion to secure favorable access to natural resources in Latin America, lobbied for access to international stations for zoological and oceanographic study, promoted efforts to conserve natural resources and sought government aid to protect their intellectual property rights; see for example Burstyn 1980, and Bugos and Kevles 1992. I thank Carl-Henry Geschwind for helpful discussions on this point.
4 Anna K. Nelson, session commentary, Cold War Science and Foreign Policy session, History of Science Society meeting, Santa Fe, New Mexico, October 1993.

5 Struik 1991, 79–80 and Schroeder-Gudehus 1990, 909; a penetrating analysis is found in Daston 1990.

6 Gillispie 1992; see also Gillispie 1983, Howse 1989, Herrmann 1984, Edgarton 1990, and Hobsbawm 1989, 243–261.

7 DeVorkin 1981 and Friedman 1990. On the work of the International Education Board, see Kohler 1991, 133–61.

8 Kevles 1970, Schroeder-Gudehus 1973, and Schroeder-Gudehus 1990; see also Guber 1975.

9 See, e.g., Alperovitz 1965, Goldberg 1992, Goldberg (forthcoming), Herken 1988, Holloway 1994, Rhodes 1986, Sherwin 1977, Walker 1990, and Weart 1989.

10 The post-war successor to the Office of Scientific Research and Development (OSRD), initially termed the Joint Research and Development Board (JRDB).

11 Thackray 1992, viii.

12 Geiger 1993, Dupree 1980, Needell 1993, DeVorkin 1992, Hershberg 1993, Leslie 1993, Lowen (1997), and Smith *et al.* 1989.

13 Herken 1992 and Kevles 1992.

14 Dennis 1994 and Wang 1992.

15 Crawford, Shinn, and Sörlin 1992, Elzinga 1993, Krige and Sebesta 1994, and Rabkin 1988; see also Bulkeley 1991, Burrows 1986, Jones 1988, McDougall 1985, Logsdon 1970, Logsdon and Dupas 1994, Launius 1994, and Allan A. Needell and R.E. Doel, "Science in the national interest: scientists and the state department, 1945–1960" (in preparation).

16 These include Hewlett and Duncan 1969, and Hewlett and Holl 1989. Hewlett and Holl explore the broader contexts of this agency more than earlier works in this series, and provide a discussion of international science on pp. 530–536, but their focus nevertheless remains national developments. Technical issues are emphasized in a recent work by Hoddeson *et al.* of Los Alamos during the Oppenheimer years, 1943–45 (in Judson and Söderqvist, forthcoming). An excellent historiographic overview is Seidel 1990.

17 Recent work by Silvan S. Schweber on Ford Foundation funding for Western European physics in the 1950s indicates how rich this topic promises to be; see Schweber 1994.

18 See, e.g., Gaddis 1982, Leffler 1992, and Rosenberg 1986.

19 Gimbel 1990 and Richard Vernon Damms, *Scientists and Statesmen: President Eisenhower's Science Advisers and National Security Policy, 1953–1961* (Ph.D. dissertation, Ohio State University, 1993); on corporatism (also termed associationalism), see Hogan 1990, Hogan 1987, and Rosenberg 1982.

20 Quoted in Leffler 1990, 143; see also Gaddis 1987, Divine 1965, Divine 1993, Smith 1983, Skolnikoff 1967, Laqueur 1993, 54–68, and Joseph A. Manzione, *The American Scientific Community, the United States Government, and the Issue of International Scientific Relations during the Cold War* (Ph.D. dissertation, University of Michigan, 1992).

21 Hunt 1990, Hunt 1987, Ezrahi 1990, and Weart 1989.

22 Gelb and Betts 1979, Gary 1991.

23 Darling 1990 and Montague 1992. Gaps in these histories remain substantial, as neither history for example reports such psychological drug testing carried out by CIA researchers in the early 1950s; see Marks 1979 and, for additional background, *Final Report* 1976. Little information on scientific intelligence appears in the standard historical account of this agency (Jeffreys-Jones 1989).

24 Berkner 1950. This issue has been analyzed in Needell (1996); see also Needell and Doel, "Science in the national interest" *op. cit.*, and Needell (forthcoming).

25 Nelkin 1972 and Smith 1992.

26 Sullivan 1961 and Bullis 1973.

27 McDougall 1985, Needell (1996), Needell (forthcoming).

28 Quoted in Gaddis 1992, 94; see also Winks 1988.

29 G.C. Marshall to President Truman, June 3, 1948 (Box 159, President's Secretary's Files, HST); Report of NSC on Position of US with Respect to Scandinavia, Dec. 16, 1948 (NSC Records, HST); see also Kennedy-Minott 1990, 5–11 and Leffler 1992, 211–218.

30 "an estimate of Swedish capabilities in science", Aug. 9, 1949 (Box 257, President's Secretary's Files, Truman Library). On the accuracy of report, see Aant Elzinga to author, personal communication; an intriguing omission in this report is geophysics, covered by other intelligence agencies.

31 Lasby 1971; see also Neufeld 1994, 267–273.

32 The often-cited Lilly 1968 makes such a claim; a similar argument concerning the Operations Coordinating Board is found in Jeffreys-Jones 1989, 121. While John Prados notes that PSB became the largest component of NSC operations in 1991, he too notes bureaucratic entanglements which limited its effectiveness; see Prados 1991, 52–56.

33 Bulkeley 1991 and Martin J. Medhurst, "Atoms for peace and nuclear hegemony: the rhetorical structure of a Cold War campaign", Presentation to the Western States Communication Association, Feb. 25, 1994, San Jose, California, in author's possession. On psychological warfare, see Needell 1993.

34 James Conant to Roger Adams, Sept. 19, 1946 (Box 7, Roger Adams papers, University of Illinois). Critical studies include Bernstein 1988 and Hershberg 1993, esp. 365–367, and Wright 1990; still extremely valuable are Hersh 1968 and Moon 1984. Vladislav M. Zubok recently reported on a special KGB division charged with fabricating disinformation about U.S. use of bio-chemical agents; see Zubok 1994, especially p. 25.

35 Existing environmental histories, such as Hays 1987, have tended to focus on the domestic roots of policy for the 1960s; one exception is Roderick Nash's "International Perspective" chapter (in Nash 1982). Thomas G. Paterson has underscored the need for environmental perspectives for diplomatic history

in Paterson 1991. An important recent study is Stuart W. Leslie, "Beyond point IV: American technology transfer programs in the Cold War", History of Science Society Annual Meeting, Oct. 1994.

36 Ninkovich 1981, 167.

37 For example, Killian 1977 and Kistiakowsky 1976. I thank Jefferson P. Marquis and Martin Collins for helpful discussions on this point.

38 I base these estimates on extensive work in the Truman and Eisenhower presidential libraries during July and August 1994. Historians interested in pursuing FBI records can profit by reading the introduction to Diamond 1992.

39 George B. Kistiakowsky Papers, Pusey Archives, Harvard University, and Frank Press papers, Archives and Special Collections, MIT.

40 Papers of Roger Adams, University of Illinois archives, Champaign Urbana, Illinois. Because of the paucity of other materials, the Adams papers nevertheless remain an important source for understanding Allied interest in German science in the postwar period. An important new work at the juncture of diplomatic history and history of science which draws on the Adams papers is Cassidy 1994.

41 In these instances, material is not awaiting declassification; it is either missing, returned to originating agencies, destroyed, or still to be donated to the archives by surviving family members.

42 Gerard P. Kuiper papers, University of Arizona library, Tucson; on Kuiper's relationship to the CIA, see Doel 1992 and Doel (forthcoming-1).

43 Frank Press collection, Special Collections, Massachusetts Institute of Technology, Cambridge, Massachusetts. The Nuclear Test Ban Treaty represented one of the most significant science-based cooperative agreements between the United States and the Soviet Union in the 1960s; Kai-Hendrik Barth, a graduate student at the University of Minnesota, is currently working on this topic.

44 For instance, the financier and administrator Ferdinand Eberstadt personally wrote the report on foreign scientific intelligence, as leader of the National Security Organization committee of the 1948 Hoover Commission on the Organization of the Executive Branch of Government. Nevertheless, his otherwise large and valuable collection has virtually no material on this topic, and many original file folders apparently arrived empty at Princeton University's Seeley G. Mudd Library. Not surprisingly, little information on international science appears in the papers of Allen Dulles, director of Central Intelligence during the Eisenhower Administration, also secured by Princeton as part of its Public Policy papers series, which also includes collections for James V. Forrestal, George Kennan, and Adlai Stevenson. Given existing constraints on documents of this kind, it is to Princeton's credit that even these materials were preserved.

45 Scientists who joined the CIA, in contrast to industrial scientists, often ceased entirely to publish work in declassified journals; see for instance entries for H. Marshall Chadwell in *Science Citation Index* after he became

head of the CIA's Office of Scientific Intelligence in 1950. A classified CIA study made a similar point regarding scientific publications in the Soviet Union; see "Abstracting services as an intelligence tool for assessing Soviet chemical research", CIA/OSI 4/49, Dec. 19, 1949 (Box 257, President's Secretary's Files, HST).

46 Cross-reference sheet dated Aug. 8, 1948, re. I.N.P. Stokes, Department of Commerce (Box [CIA], Confidential File series, Truman Presidential Library); further analysis appears in Needell and Doel, *op. cit.*

47 On the last point, see Forman 1991. Transcripts of recent oral history interviews that I and other colleagues have led with scientists concerned with foreign policy issues makes clear my (and others') trepidations about probing sensitive topics at length, for fear of curtailing otherwise enlightening interviews; the problem ought not be minimized. A useful discussion of related issues appears in Lillian Hoddeson, "The conflict of memories and documents: dilemmas and pragmatics of oral history", in Judson and Söderqvist (forthcoming).

48 For instance, see the papers of Senator H. Alexander Smith, Princeton University and those of Senator Henry Jackson at the University of Washington, Seattle. An important introduction to the work of one Congressional science committee is Hechler 1980.

49 Discussion with Dennis Bilger, senior archivist, Truman Presidential Library, summer 1994.

50 Judson and Söderqvist (forthcoming).

51 Doel (forthcoming-2).

52 Beckmann Institute Archives, California Institute of Technology, Pasadena, CA.

53 Interview with Frederick Seitz (Allan A. Needell and Ronald E. Doel, July 19, 1994), and interview with Richard T. Arnold (Ronald E. Doel, Aug. 10, 1994, Center for History of Physics, American Institute of Physics, in process). Needell and I are grateful for assistance from the Center for History of Physics and the Smithsonian Institution for help in processing these interviews.

54 Paterson 1991, Pérez 1990, Palladino and Worboys 1993, and Hoch 1994.

55 Klaus-Harro Tiemann, "Die Entwicklung der sowjetisch-deutschen Wissenschaftsbeziehungen auf dem Gebiet der Geo- und Kosmoswissenschaften", 1988 manuscript in author's possession, and Krementsov 1995. Important overviews of recent work on the history of Soviet science include Alexandrov and Krementsov 1989 and Daniel Alexandrov, "Historical anthropology of science in Russia", contribution to New Trends in the History of Science and Technology seminar, St. Petersburg, June 22–23, 1994.

56 Historians should also not overlook the possibility of organizing conferences or symposia that include scientists central to international science activities, particularly when counterparts from other nations can simultaneously engage in discussions. A conference on "Astronomy and the State", featuring U.S. and Russian participants, was held in Washington, D.C. in January 1994, with

partial sponsorship from the Smithsonian Institution, IREX, and the Historical Astronomy Division of the American Astronomical Society; articles based on this session appeared in the *Journal for History of Astronomy*, vol. 26, pt 4 (Nov. 1995), Robert A. McCutcheon *et al.* eds.

57 Sakharov 1991, Holloway 1994, 367–368, and Weart and Szilard 1978.

58 On Robertson, see Robertson to Col. C.P. Nicholas, Aug. 28, 1957 (Box 4.4, Robertson collection, Caltech); on Brode, see Needell and Doel (forthcoming). David H. DeVorkin's study of the post-1945 rise of space science in the United States has illuminates the careers of several individuals proud to work on government projects, including the University of Michigan researcher William G. Dow; see DeVorkin 1992, 311.

59 Weart 1979, Bensaude-Vincent 1987, and Graham 1993.

60 Needell 1993; see also Needell 1992.

61 Rebecca Lowen is currently studying this foundation; see also Blaauw 1994, Pisani 1991, 48–49, and Needell and Doel "Science in the national interest" (*op. cit.*).

62 Record of the 468th meeting of the National Security Council, held Dec. 1, 1960 (Box 13, NSC series, Ann Whitman file, Eisenhower Presidential Library). David Holloway similarly concluded that in the U.S.S.R. "[t]he scientific community — and especially the physics community — was, for all its failings, the closest thing to civil society in the Stalinist regime"; see Holloway 1994, 363. For U.S. policy on funding science in Western Europe, see Needell (1996).

63 Quoted from Packenham 1973, 43; see also Gendzier 1985 and Rabe 1993.

64 As Marcus Cueto has demonstrated, U.S. scientists had long-standing relationships with Latin and South American scientists through the Rockefeller Foundation, which continued to fund research in these regions after 1945; see Cueto 1994 as well as Glick 1994 and Kohler 1991. Ties between the U.S. and sub-Sahara Africa were previously weaker, largely because scientists had not tried to compete with existing imperial networks; see the "trilogy" by Lewis Pyenson (Pyenson 1985, Pyenson 1989, and Pyenson 1993) as well as Worthington 1958. Neither of the standard histories of the National Science Foundation, Lomask 1976 and England 1983, discuss the Foundation's involvement in international science.

65 IGY programs have been cited as the inspiration of NSF interest in the Americas, but foundation leaders expressed a desire to promote Latin American science as early as 1954, before the IGY was fully formulated and had gained NSF funding; see Testimony of Dr. Frederick M. Bernthal, March 17, 1992, in *Latin American Scientific Cooperation: Joint Hearing Before the Subcommittee on Science of the Committee on Science, Space, and Technology, U.S. House of Representatives, One Hundredth Congress, Second Session* (March 17, 1992), 25; see also Alan Waterman, Diary Note, July 27, 1954 (Box 10, NSF record series), and E.B. Skolnikoff to James Killian, memo on Science and Foreign Relations Panel Meeting, Oct. 23,

1958, President's Science Advisory Committee files (Box 111, Record Group 359, both National Archives, Washington, D.C.). On the ideological dimensions of U.S.-Latin American relations, see LaFeber 1993 and Smith 1994.

66 Packenham 1973 and Hogan 1987.

67 The standard work on the political activism of scientists in the 1930s is Kuznick 1987; see also Stuart W. Leslie, "Beyond point IV: American technology transfer programs in the Cold War", History of Science Society Annual Meeting, New Orleans, October 1994.

68 Dennis 1994 and Leslie 1993, 255–256.

69 Needell and Doel, "Science in the national interest" (*op. cit.*).

70 Mukerji 1989, 21.

71 "Soviet scientific and technical manpower", classified secret report, n.d. [circa June 1953] (Box 3, White House, NSC Staff, NSC Registry Series, Eisenhower Presidential Library, Abilene, Kansas).

72 Forman 1987.

73 See for instance Jones 1988, Graham 1973, Cassidy 1994, and Home and Low 1993.

74 Cabral 1994 and Elzinga 1993.

75 I am currently writing a book on the rise and institutionalization of academic geophysics in North America, 1920–1970, from which this example derives; for sources, see, e.g., J. Tuzo Wilson, "Annual Report for 1953: Department of Physics, Geophysics Laboratory", draft, n.d. [circa 154] (Box 23, J.T. Wilson papers, University of Toronto), author's interview with J. Tuzo Wilson February 1993, and author's interview with George Garland February 1993, both at Center for History of Physics, American Institute of Physics, College Park, MD.

76 Merton 1973; see also Daston 1990, 114. I am grateful to Frederic L. Holmes for fruitful discussions on this point.

77 Pisani 1991, 4; a general review of this issue is Berkowitz and Goodman, 1989.

78 See, e.g., David Z. Beckler to Ralph Clark, Dec. 2, 1947 (Box 1, NSC records, Truman library), and "Collected Abstracts of Committee Minutes for [Secretary of War Robert S.] Paterson", Eberstadt Committee records (Box 73, Eberstadt papers, Princeton University); see also Laqeuer 1993. A new work which addresses critical aspects of this issue, limited to nuclear intelligence, is Ziegler and Jacobson 1995; I thank Ziegler for sharing a manuscript version with me.

79 This discussion derives from selected biographical compilations of scientists in the CIA; the importance that academic chemists attached to applied research can be found in Roger Adams to Warren Weaver, March 17, 1959 (Box 7, Adams papers, University of Illinois).

80 Palladino and Worboys 1993.

81 Rosenberg 1988, 569–570.

BIBLIOGRAPHY

Alexandrov, Daniel and Nikolai Krementsov, "Opyt putevoditelia po heizvedannoi zemle: predvaritelnyi ocherk sotsialnoi istorii sovetskoi nauki" [Experiences of a guide through an unexplored land: preliminary sketch of the social history of Soviet science]", *Voprosy istorii estestvoznaniia i tekhniki* [Issues in the History of Natural Science and Technology], vol. 4 (1989), 67–87.

Alperovitz, Gar, *Atomic Diplomacy: Hiroshima and Potsdam* (New York: Prager, 1965).

Bensaude-Vincent, Bernadetta, *Langevin: science et vigilance* (Paris: Belin, 1987).

Berkner, Lloyd V., *Science and Foreign Relations (Report as Special Consultant to the Secretary of State on Survey of Department Responsibilities in International Science* (Washington, D.C., 1950).

Berkowitz, Bruce D. and Alan Goodman, *Strategic Intelligence for American National Security* (Princeton: Princeton University Press, 1989).

Bernstein, Barton J., "America's biological warfare program in the Second World War", *Journal of Strategic Studies*, vol. 11 (1988), 292–317.

Blaauw, Adriaan, *History of the IAU: The Birth and First Half-Century of the International Astronomical Union* (New York: Kluwer, 1994).

Bugos, Glenn E. and Daniel J. Kevles, "Plants as intellectual property: American practice, law, and policy in world context", *Osiris*, vol. 7 (1992), 75–104.

Bulkeley, Rip, *The Sputniks Crisis and Early United States Space Policy* (Bloomington: Indiana University Press, 1991).

Bullis, Harold, *The Political Legacy of the International Geophysical Year* (Washington, D.C.: Government Printing Office, 1973).

Burrows, William E., *Deep Black: Space Espionage and National Security* (New York: Random House, 1986).

Burstyn, Harold L., "Reviving American oceanography: Frank Lillie, Wickliffe Rose, and the founding of the Woods Hole Oceanographic Institution", pp. 57–66 in Mary Sears and Daniel Merriman (eds.), *Oceanography: The Past* (New York: Springer, 1980).

Cabral, Regis, *The Brazilian Nuclear Debate, 1945–1955* (Göteborg: Science, Technology, Ideology, Culture Series No. 8, Göteborg University, 1994).

Cassidy, David, "Controlling German science, I: U.S. and allied forces in Germany, 1945–1947", *Historical Studies in the Physical and Biological Sciences*, vol. 24 (1994), 197–235.

Crawford, Elisabeth, Terry Shinn, and Sverker Sörlin (eds.), *Denationalizing Science: The Contexts of International Scientific Practice* (Dordrecht: Kluwer, 1992).

Cueto, Marcus, "Laboratory styles in Argentine physiology", *Isis*, vol. 85 (1994), 228–246.

Darling, Arthur B., *The Central Intelligence Agency: An Instrument of Government, to 1950* (University Park, PA.: Pennsylvania State University Press, 1990).

Daston, Lorraine, "Nationalism and scientific neutrality under Napoleon", pp. 95–119 in Tore Frängsmyr (ed.), *Solomon's House Revisited: The Organization and Institutionalization of Science* (Canton, MA: Science History Publications, 1990).

Dennis, Michael Aaron, " 'Our first line of defense': two university laboratories in the postwar American state", *Isis*, vol. 85 (1994), 427–455.

Diamond, Sigmund, *Compromised Campus: The Collaboration of Universities with the Intelligence Community, 1945–1955* (New York: Oxford University Press, 1992).

Divine, Robert A., *Blowing on the Wind: The Nuclear Test Ban Debate, 1956–1961* (New York: Doubleday, 1965).

Divine, Robert A., *The Sputnik Challenge* (New York: Doubleday, 1993).

DeVorkin, David H., "Community and spectral classification in astrophysics: the acceptance of E.C. Pickering's system in 1910", *Isis*, vol. 72 (1981), 29–49.

DeVorkin, David H., *Science with a Vengeance: The Military Origins of Space Science* (New York: Springer, 1992).

Doel, Ronald E., "Evaluating Soviet lunar science in Cold War America", *Osiris*, vol. 7 (1992), 238–264.

Doel, Ronald E., "Documents and reports: G.P. Kuiper's report on Soviet astronomy to the CIA", *Istoriko-astronomicheskie issledovanniia* (forthcoming-1).

Doel, Ronald E., "Oral history, archival sources, and the historiography of modern science", in *Interviews in Writing the History of Recent Science*, Horace F. Judson and Thomas Söderqvist, eds. (Cambridge, MA: Harvard University Press, forthcoming-2).

Dorsey, Jurk, "Scientists, citizens, and statesmen: US-Canadian wildlife protection treaties in the progressive era", *Diplomatic History*, vol. 19 (1995), 407–430.

Drew, Sidney, D. and Sergei, P. Kapilźa (eds.), *Sakharov Remembered: A Tribute by Friends and Colleagues* (New York: American Institute of Physics, 1991).

Dupree, A. Hunter, "The great instauration of 1940: the organization of scientific research for war", pp. 443–467 in Gerald Holton (ed.), *The Twentieth Century Sciences: Studies in the Biography of Ideas* (New York: W.W. Norton, 1972).

Dupree, A. Hunter, "A historian's view of advice to the president on science: retrospect and prescript", *Technology in Society*, vol. 2 (1980), 175–190.

Edgarton, D.E.M., "Science and war", pp. 920–933 in R.C. Olby *et al.* (eds.), *Companion to the History of Science* (London: Routledge, 1990).

Elzinga, Aant, "Antarctica: the construction of a continent by and for science", pp. 73–106 in Elisabeth Crawford, Terry Shinn, and Sverker Sörlin (eds.), *Denationalizing Science: The Contexts of International Scientific Practice* (Dordrecht: Kluwer, 1993).

England, J. Merton, *A Patron for Pure Science: The National Science Foundation's Formative Years, 1945–57* (Washington, D.C.: National Science Foundation, 1983).

Ezrahi, Yaron, *The Descent of Icarus: Science and the Transformation of Contemporary Democracy* (Cambridge, MA.: Harvard University Press, 1990).

Final Report of the Select Committee to Study Governmental Operations with Respect to Intelligence Activities, U.S. Senate, No. 94–755 (Washington: Government Printing Office, 1976).

Forman, Paul, "Behind quantum electronics: natural security as basis for physical research in the United States, 1940–1960", *Historical Studies in the Physical and Biological Sciences*, vol. 18 (1987), 149–229.

Forman, Paul, "Independence, not transcendence, for the historian of science", *Isis*, vol. 82 (1991), 71–86.

Friedman, Robert Marc, "The Nobel Prizes and the invigoration of Swedish science: some considerations", pp. 193–207 in Tore Frängsmyr (ed.), *Solomon's House Revisited: The Organization and Institutionalization of Science* (Canton, MA: Science History Publications, 1990).

Gaddis, John Lewis, *Strategies of Containment: A Critical Appraisal of Postwar American National Security Policy* (New York: Oxford University 1982).

Gaddis, John Lewis, "Learning to live with transparency: the emergence of a reconnaissance satellite regime", pp. 195–214 in John Lewis Gaddis (ed.), *The Long Peace: Inquiries into the History of the Cold War* (New York: Oxford University Press, 1987).

Gaddis, John Lewis, *The United States and the End of the Cold War* (New York: Oxford University Press, 1992).

Gary, J. Clifford, "Bureaucratic politics", pp. 141–150 in Michael J. Hogan and Thomas G. Paterson (eds.), *Explaining the History of American Foreign Relations* (New York: Cambridge University Press, 1991).

Geiger, Roger L., *Research and Relevant Knowledge: American Research Universities Since World War II* (New York: Oxford University Press, 1993).

Gelb, Leslie H. (with Richard K. Betts), *The Irony of Vietnam: The System Worked* (Washington, D.C.: Brookings Institution, 1979).

Gendzier, Irene L., *Managing Political Change: Social Scientists and the Third World* (Boulder, CO.: Westview Press, 1985).

Gillispie, Charles C., *The Professionalization of Science: France, 1770–1830, Compared to the United States, 1910–1970* (Kyoto: Doshisha University Press, 1983).

Gillispie, Charles Coulston, "Science and secret weapons development in revolutionary France, 1792–1804: a documentary history", *Historical Studies in the Physical and Biological Sciences*, vol. 23 (1992), 35–152.

Gimbel, John, *Science, Technology, and Reparations: Exploitation and Plunder in Postwar Germany* (Stanford: Stanford University Press, 1990).

Glick, Thomas, "The Rockefeller Foundation and the emergence of genetics in Brazil, 1943–1960", pp. 154–169 in Marcus Cueto (ed.), *Missionaries of Science: The Rockefeller Foundation and Latin America* (Bloomington: Indiana University Press, 1994).

Goldberg, Stanley, "Inventing a climate of opinion: Vannevar Bush and the decision to build the bomb", *Isis*, vol. 83 (1992), 429–452.

Goldberg, Stanley, "General Groves and the atomic West: the making and meaning of Hanford", in Bruce Hevly and John Findlay, eds., *The Atomic West*, (Seattle: University of Washington Press, forthcoming).

Graham, Loren R., *Science in Russia and the Soviet Union: A Short History* (New York: Cambridge University Press, 1993).

Guber, Carol S., *Mars and Minerva: World War I and the Uses of the Higher Learning in America* (Baton Rouge: Louisiana State University Press, 1975).

Hays, Samuel P., *Beauty, Health, and Permanence: Environmental Politics in the United States, 1955–1985* (New York: Cambridge University Press, 1987).

Hechler, Ken, *Toward the Endless Frontier: History of the Committee on Science and Technology, 1959–79* (Washington, D.C.: Government Printing Office, 1980).

Herken, Gregg, *The Winning Weapon: The Atomic Bomb in the Cold War, 1945–1950* (Princeton, N.J.: Princeton University Press, 1988).

Herken, Gregg, *Cardinal Choices: Presidential Science Advising from the Atomic Bomb to SDI* (New York: Oxford University Press, 1992).

Hersh, Seymour M., *Chemical and Biological Warfare: America's Hidden Arsenal* (Indianapolis and New York: Bobbs-Merrill, 1968).

Hershberg, James, *James B. Conant: Harvard to Hiroshima and the Making of the Nuclear Age* (New York: Knopf, 1993).

Herrmann, Dieter B., *The History of Astronomy from Herschel to Hertzsprung* (revised, transl. by Kevin Krisciunas, New York: Cambridge University Press, 1984).

Hewlett, Richard G. and Francis Duncan, *Atomic Shield, 1947–1952: A History of the United States Atomic Energy Commission* (University Park: Pennsylvania State University Press, 1969).

Hewlett, Richard G. and Jack M. Holl, *Atoms for Peace and War, 1953–1961: Eisenhower and the Atomic Energy Commission* (Berkeley: University of California Press, 1989).

Hobsbawm, Eric J., *Age of Empire, 1875–1914* (New York: Vintage Books, 1989).

Hoch, Paul, "[Essay review:] Whose scientific internationalism?", *British Journal for the History of Science*, vol. 27 (1994), 345–349.

Hoddeson, Lillian, "Dilemmas and pragmatics in oral history", in *Interviews in Writing the History of Recent Science*, Horace F. Judson and Thomas Söderqvist, eds. (Cambridge, MA: Harvard University Press, forthcoming).

Hogan, Michael J., *The Marshall Plan: America, Britain, and the Reconstruction of Western Europe, 1947–1952* (New York: Cambridge University Press, 1987).

Hogan, Michael J., "Corporatism", *Journal of American History*, vol. 77 (1990), 153–160.

Holloway, David, *Stalin and the Bomb: The Soviet Union and Atomic Energy, 1939–1956* (New Haven: Yale University Press, 1994).

Home, R.W. and Morris F. Low, "Postwar scientific intelligence missions to Japan", *Isis*, vol. 84 (1993), 527–537.

Howse, Derek, *Nevil Maskelyne: The Seaman's Astronomer* (New York: Cambridge University Press, 1989).

Hunt, Michael H., "Ideology", *Journal of American History*, vol. 77 (1990), 108–115.

Hunt, Michael H., *Ideology and U.S. Foreign Policy* (New Haven: Yale University Press, 1987).

Jeffreys-Jones, Rhodri, *The CIA and American Democracy* (New Haven: Yale University Press, 1989).

Jones, Greta, *Science, Politics, and the Cold War* (London: Routledge, 1988).

Judson, Horace F. and Thomas Söderqvist (eds.), *Interviews in Writing the History of Recent Science* (Cambridge, MA: Harvard University Press, forthcoming).

Kennedy-Minott, Rodney, *Lonely Path to Follow: Non-Aligned Sweden, United States/NATO, and the U.S.S.R.* (Stanford: Hoover Institution, 1990).

Kevles, Daniel J., " 'Into hostile political camps': the reorganization of international science in World War I", *Isis*, vol. 62 (1970), 47–60.

Kevles, Daniel J., "K_1S_2: Korean, science and the state", pp. 312–333 in Peter Galison and Bruce Hevly (eds.) *Big Science: The Growth of Large-Scale Research* (Stanford: Stanford University Press, 1992).

Killian, James R., *Sputnik, Scientists, and Eisenhower: Memoirs of the First Special Assistant to the President for Science and Technology* (Cambridge, MA: MIT Press, 1977).

Kistiakowsky, George B., *A Scientist at the White House: The Private Diary of President Eisenhower's Special Assistant for Science and Technology* (Cambridge, MA: Harvard University Press, 1976).

Kohler, Robert E., *Partners in Science: Foundations and Natural Scientists, 1900–1945* (Chicago: University of Chicago Press, 1991).

Krementsov, Nikolai, "Soviet science on the threshold of the Cold War: the KR affair", *Journal for the History and Philosophy of the Life Sciences*, vol. 17 (1995), 419–446.

Krige, John and Lorenza Sebesta, "US-European cooperation in space in the decade after Sputnik", pp. 263–286 in Giuliana Gemelli (ed.), *Big Culture: Intellectual Co-operation in Large-Scale Cultural and Technical Systems: An Historical Approach* (Bologna: Editrice, 1994).

Kuznick, Peter J., *Beyond the Laboratory: Scientists as Political Activists in 1930s America* (Chicago: University of Chicago Press, 1987).

LaFeber, Walter, *Inevitable Revolutions: The United States in Central America* (2nd ed. New York: W.W. Norton, 1993).

Laqueur, Walter, *The Uses and Limits of Intelligence* (New Brunswick, NJ: Transaction Publishers, 1993).

Lasby, Clarence G., *Project Paperclip: German Scientists and the Cold War* (New York: Atheneum, 1971).

Launius, Roger, NASA, *A History of the US Civil Space Program* (Malabar, FL: Krieger, 1994).

Leffler, Melvyn P., "National security", *Journal of American History*, vol. 77 (1990), 143–152.

Leffler, Melvyn P., *A Preponderance of Power: National Security, the Truman Administration, and the Cold War* (Stanford: Stanford University Press, 1992).

Leslie, Stuart W., *The Cold War and American Science* (New York: Columbia University Press, 1993).

Lilly, Edward P., "The Psychological Strategy Board and its predecessors: foreign policy coordination", pp. 337–382 in Gartano L. Vincitorio (ed.), *Studies in Modern History* (New York: St. John's University Press, 1968).

Logsdon, John M., *The Decision to Go to the Moon: Project Apollo and the National Interest* (Cambridge, MA.: MIT Press, 1970).

Logsdon, John M. and Alain Dupas, "Was the race to the moon real?", *Scientific American*, vol. 270, 6 (June 1994), 36–43.

Lomask, Milton, *A Minor Miracle: An Informal History of the National Science Foundation* (Washington, D.C.: National Science Foundation, 1976).

Lowen, Rebecca, *Creating the Cold War University: The Transformation of Stanford* (Los Angeles: University of California Press, 1997).

McDougall, Walter, *The Heavens and the Earth: A Political History of the Space Age* (New York: Basic Books, 1985).

Marks, John D., *The Search for The 'Manchurian Candidate'* (New York: Times Books, 1979).

Merton, Robert K., *The Sociology of Science: Theoretical and Empirical Investigations* (Chicago: University of Chicago Press, 1973).

Moon, John Ellis van Courtland, "Chemical weapons and deterrence: The World War II experience", *International Security*, vol. 8 (1984), 3–35.

Montague, Ludwell Lee, *General Walter Bedell Smith as Director of Central Intelligence, October 1950–February 1953* (University Park, PA: Pennsylvania State University Press, 1992).

Mukerji, Chandra, *A Fragile Power: Scientists and the State* (Princeton: Princeton University Press, 1989).

Nash, Roderick, *Wilderness and the American Mind* (3rd ed., New Haven: Yale University Press, 1982).

Needell, Allan A., "From military research to Big Science: Lloyd Berkner and science-statesmanship in the postwar era", pp. 290–311 in Peter Galison and Bruce Hevly (eds.), *Big Science: The Growth of Large-Scale Research* (Stanford: Stanford University Press, 1992).

Needell, Allan A., "Truth is my weapon: project TROY, political warfare, and government-academic relations in the national security state", *Diplomatic History*, vol. 17 (1993), 399–420.

Needell, Allan A., *Over the Horizon: Lloyd V. Berkner, The Technocratic Vision, and the Cold War* (forthcoming-1).

Needell, Allan A., "Rabi, Berkner, and the rehabilitation of science in Europe: the Cold War context of American support for international science 1945–1958", pp. 289–305 in Francis H. Heller and John Gillingham (eds.), *The United States and Integration of Europe* (New York: St. Martin's Press, 1996).

Nelkin, Dorothy, *The University and Military Research: Moral Politics at M.I.T.* (Ithaca: Cornell University Press, 1972).

Neufeld, Michael J., *The Rocket and the Reich: Peenemünde and the Coming of the Ballistic Missile Era* (New York: Free Press, 1994).

Ninkovich, Frank, *The Diplomacy of Ideas: U.S. Foreign Policy and Cultural Relations, 1938–1950* (New York: Cambridge University Press, 1981).

Packenham, Robert A., *Liberal America and the Third World* (Princeton: Princeton University Press, 1973).

Palladino, Paolo and Michael Worboys, "Science and imperialism", *Isis*, vol. 84 (1993), 91–102.

Paterson, Thomas, "Defining and doing the history of American foreign relations: a primer", pp. 36–54 in Michael J. Hogan and Thomas G. Paterson (eds.), *Explaining the History of American Foreign Relations* (New York: Cambridge University Press, 1991).

Pérez, Louis A., Jr., "Dependency", *Journal of American History*, vol. 77 (1990), 133–142.

Pisani, Sallie, *The CIA and the Marshall Plan* (Lawrence, KA: University Press of Kansas, 1991).

Prados, John, *Keepers of the Keys: A History of the National Security Council from Truman to Bush* (New York: William Morrow, 1991).

Pyenson, Lewis, *Cultural Imperialism and Exact Sciences: German Expansion Overseas, 1900–1930* (New York: P. Lang, 1985).

Pyenson, Lewis, *Empire of Reason: Exact Sciences in Indonesia, 1840–1940* (Leiden: E.J. Brill, 1989).

Pyenson, Lewis, *Civilizing Mission: Exact Sciences and French Overseas Expansion, 1830–1940* (Baltimore: Johns Hopkins University Press, 1993).

Rabe, Stephen G., "Eisenhower revisionism: a decade of scholarship", *Diplomatic History*, vol. 17 (1993), 97–115.

Rabkin, Yakov M., *Science between the Superpowers* (New York: Priority Press, 1988).

Rhodes, Richard, *The Making of the Atomic Bomb* (New York: Simon and Schuster, 1986).

Rosenberg, Charles E., "Woods or trees: ideas and actors in the history of science", *Isis*, vol. 79 (1988), 565–570.

Rosenberg, David Alan, "The origins of overkill: nuclear weapons and American strategy", pp. 123–195 in Norman A. Graebner (ed.), *The National Security: Its Theory and Practice, 1945–1960* (New York: Oxford University Press, 1986).

Rosenberg, Emily S., *Spreading the American Dream: American Economic and Cultural Expansion, 1890–1945* (New York: Hill and Wang, 1982).

Schröder-Gudehus, Brigitte, "Challenge to transnational loyalties: international scientific organizations after the First World War", *Science Studies*, vol. 3 (1973), 93–118.

Schroeder-Gudehus, Brigitte, "Nationalism and internationalism", pp. 909–919 in R.C. Olby *et al.* (eds.), *Companion to the History of Modern Science* (London: Routledge, 1990).

Schweber, Silvan S., "Theoretical physics and the restructuring of the physical sciences: 1925–1975", pp. 131–162 in Giuliana Gemelli (ed.), *Big Culture: Intellectual Cooperation in Large-Scale Cultural and Technical Systems. An Historical Approach* (Bologna: Editrice, 1994).

Seidel, Robert W., "Clio and the complex: recent historiography of science and national security", *Proceedings of the American Philosophical Society*, vol. 134 (1990), 420–441.

Sherwin, Martin J., *A World Destroyed: The Atomic Bomb and the Grand Allegiance* (New York: Vintage Books, 1977).

Skolnikoff, Eugene B., *Science, Technology, and American Foreign Policy* (Cambridge: MIT Press, 1967).

Smith, Bruce L.R., *The Advisors: Scientists in the Policy Process* (Washington, D.C.: Brookings Institution, 1992).

Smith, Gaddis, *The Last Years of the Monroe Doctrine 1945–1993* (New York: Hill and Wang, 1994).

Smith, Michael L., "Selling the moon: The U.S. manned space program and the triumph of commodity scientism", pp. 175–209 in R.W. Fox and T.J. Jackson Lears (eds.), *The Culture of Consumption* (New York: Pantheon, 1983).

Smith, Robert W. *et al.*, *The Space Telescope: A Study of NASA, Science, Technology, and Politics* (New York: Cambridge University Press, 1989).

Struik, Dirk, *Yankee Science in the Making: Science and Engineering in New England from Colonial Times to the Civil War* (New York: Dover, 1991).

Sullivan, Walter, *Assault on the Unknown: The International Geophysical Year* (New York: McGraw-Hill, 1961).

Thackray, Arnold, "Preface" [to A. Thackray, ed., Science After Forty], *Osiris*, vol. 7 (1992), vi–ix.

Walker, J. Samuel, "The Decision to Use the Bomb: An Historiographic Update", *Diplomatic History*, vol. 14 (1990), 75–91.

Wang, Jessica, "Science, security, and the Cold War: the case of E.U. Condon", *Isis*, vol. 83 (1992), 238–269.

Weart, Spencer R., *Scientists in Power* (Cambridge, MA: Harvard University Press, 1979).

Weart, Spencer R., *Nuclear Fear: A History of Images* (Cambridge, MA: Harvard University Press, 1989).

Weart, Spencer R. and Gertrud Weiss Szilard (eds.), *Leo Szilard, His Version of the Facts: Selected Recollections and Correspondence* (Cambridge, MA: MIT Press, 1978).

Winks, Robin, *Cloak and Gown: Scholars in America's Secret War* (New York: Quill, 1988).

Worthington, E. Barton, *Science in the Development of Africa: A Review of the Contribution of Physical and Biological Knowledge South of the Sahara* (London: Commission for Technical Co-operation in Africa South of the Sahara, 1958).

Wright, Susan (ed.), *Preventing a Biological Arms Race* (Cambridge, MA: MIT Press, 1990).

Yergin, Daniel, *The Prize: The Epic Quest for Oil, Money & Power* (New York: Simon and Schuster, 1992).

Ziegler, Charles A. and David Jacobson, *Spying Without Spies: America's Secret Nuclear Surveillance System* (Westport, CT: Praeger, 1995).

Zubok, Vladislav M., "Spy vs. spy: the KGB vs. the CIA, 1960–1962", *Cold War International History Project Bulletin* (Fall 1994), 22–33.

CHAPTER 13

Who's Afraid of the History of Contemporary Science?

Steve Fuller

INTRODUCTION

The history of contemporary science is a pursuit that challenges our taken-for-granted notions of both 'history' and 'science'. Thomas S. Kuhn is on record as holding that the pursuit is impossible: History can only be written about periods in which the historian does not have a stake in what happens because the issues of greatest concern to the historical agents no longer have the same relevance.[1] Historians enter only once the case is closed. Thus, Kuhn accepts the familiar paradox that the historian cannot represent the agents in their own terms unless she is detached from their life concerns. Although the prospect of incommensurable paradigms may wreak havoc on philosophical theories of scientific rationality, it nevertheless enables the historian to do her job.[2] However, when it comes to contemporary paradigms, the historian's role is almost exclusively that of an archivist, one who ensures that a record of today's scientists is preserved for tomorrow's historians to analyze from the requisite distance. Kuhn has taken his own historiography to heart, writing about the history of physics no later than 1912 (roughly the end of classical quantum mechanics),[3] while conducting interviews and collecting papers related to the great quantum physicists later in the century.[4]

Not surprisingly, a range of alternative views emerged from the International Workshop on the Historiography of Contemporary Science, Technology, and Medicine, held in Göteborg, Sweden, September 1994. They may be defined by their logical distance from Kuhn's position: *contradictory*, *contrary*, and *orthogonal*. In each case, the historical sensibility is guided more by the obstacles and opportunities afforded by the ultimate object of historical inquiry — the historical agent(s) — than by those afforded by its proximate object, the archived text. Indeed, by drawing attention to the distinction between historical agents and the texts that represent them, historians of the contemporary period are potentially problematizing historical scholarship as it is normally understood. Instead of a 'closed book', the proper object of historical inquiry would seem to be the 'open conversation'.

For example, Thomas Söderqvist directly contradicts Kuhn's historiographical dictum with his heavy reliance on self-conducted interviews — 200 hours with one Nobel Prize winning immunologist — as primary research data.[5] In these interviews, the historian is not merely trying to construct an archival record for establishing who did what, when, how, and why, but more importantly he is becoming immersed in the lifeworld of a scientist, or more precisely, discovering what it is like to be a scientist. Thus, the main product of the 200 hours is not a mountain of facts but bonds of intimacy and trust that enable the historian to appreciate the scientist's existential ambivalence both in its own terms and as symptomatic of multiple social forces acting upon the scientist. Perhaps the main challenge posed by this kind of historiography is the adoption of a style of writing that does justice to the phenomenology of the scientist's experience that at the same time accords proper weight to those aspects of the scientist's reality that he would rather forget or revise.

A position contrary to Kuhn's was presented by sociologist Harry Collins, who argued that the contemporary period is the *only* one about which the historian can reliably write, mainly because the scientists concerned have not yet had an opportunity to forge canonical accounts of their exploits, which then function as myths obscuring any hope of getting at really happened.[6] The social relationship that typifies the interviews Collins conducts with scientists is the formal exchange. He wants to maintain a clear separation between the interests of the scientist and the historian, thereby ensuring the autonomy of both. In that case, the historian must provide an incentive for the scientist to speak with him at the length and depth that he requires. Collins' solution is to trade on what he has learned about the other scientists he has interviewed.

A third position, one orthogonal to Kuhn's, may be instructively contrasted with Collins' as well. It was most forcefully represented at the workshop by a young historian, Jeff Hughes.[7] By way of analogy, consider the difference between inquisitorial and adversarial legal systems. If Collins' historian is like the judge who launches an independent investigation into whether a given event constitutes an infraction of the law, Hughes' is more like the judge who is forced to decide, or at least negotiate, between highly motivated litigants who see the courtroom as their ultimate site of vindication. For Hughes, the historian of contemporary science potentially faces the entire history of science as her bailiwick, as scientists attempt to protect their own legacy and undermine those of others. Thus, arguments about whether the past or the present is a more apt object of historical inquiry are beside the point, since the past is the principal resource out of which the present is constructed.

Of these three orientations to the history of contemporary science, the third comes closest to my own. But once we admit that the present is constructed from the past, we are forced to ask why scientists no longer see the history of their field as a proper part of their scientific inquiries. After all, before the First World War, most histories of science had been written by scientists who, again like judges, collected cases from the past as precedents for establishing the direction in which contemporary scientific research should go.[8] Today we normally think of this work as the prehistory of the *philosophy* of science, though at the time it would have seemed like an especially learned and public version of science policy debate. Back then there were many substantive disagreements over exactly which episodes from the history of science set the relevant precedents for future research. It is often forgotten that the people typically credited with founding the field known as 'history and philosophy of science', Auguste Comte and William Whewell, wanted to fathom the methodological secrets of physics' success more out of an interest in transferring those secrets to the backward sciences than in legitimating current trends in physics.

In contrast, today there is considerable agreement among philosophers and scientists on the basic plot structure of the history of science, with dissent arising only over how best to explain this common plot — whether it reveals a 'realist' or an 'instrumentalist' *telos*. And even this dissent is confined mostly to philosophers, since practicing scientists generally have a hard time imagining how the two *teloi* could be anything but mutually reinforcing. Moreover, while historians and sociologists manifest greater variety in the ways they understand science, it is probably not as much as had existed within the scientific community a hundred years ago, when there was still a live debate as to whether, say, the laws of Newtonian mechanics allowed for the timeframe required by Darwinian evolution — and if not, which theory would be at fault.

THE PROFESSIONALIZATION OF HISTORY OF SCIENCE AND ITS EFFECTS

Why such agreement now but not before? The answer lies in the institutionalization of 'history of science' as a field of inquiry whose domain has been insulated from contemporary science. However, the possibility of a 'history of contemporary science' has the potential for returning the field to its polemical roots. When the future is at stake, a constructivist historiography becomes suddenly appealing, if 'constructivism' is taken to mean a penchant for showing that things could have been other than they are, had events been other than they were. The availability of multiple pasts is an apt rhetorical fulcrum for leveraging into an array of alternative futures.

Whatever 'professionalism' has been gained by historians taking control of the research agenda of the history of science over the course of this century must be weighed against a subtle but real cost, namely, the overall decline in the quality of popular histories written by practicing scientists, which are still by far the ones that most non-historians read. This state-of-affairs is ironic yet non-trivial. 'Ironic' because *prima facie* history of science is a flourishing field that enjoys considerable financial and moral support from the scientific community, yet it seems to have little impact on the histories that scientists write and the public reads. 'Non-trivial' because since the end of the Cold War, popular histories have rapidly lost their traditional ability to legitimate science in the eyes of non-scientists. Consider the recent decision by the US Congress to end funding for the Supercollider, a scientific instrument whose $10 billion price-tag could not be redeemed by the claim that it would fulfill the 'historical mission' of physics.

When the history of science was still essentially contested terrain in 1900, scientists studied the historical record much more carefully than today's science popularizations seem to indicate. At the turn of the century, it was not uncommon for a Pierre Duhem or an Emile Meyerson, in the course of their philosophical histories, to draw attention to archives that would engage later generations of historians. Could it be that the scientists were gradually driven out of the archives as the historians colonized them, and that this, in turn, paved the way for the potted histories common to science popularizations and textbooks? After all, the latter phenomenon — to which Kuhn famously drew attention[9] — is more evident now than ever before. Even two or three generations ago, scientific treatises often began with fairly sophisticated philosophical histories of their topics before plunging into contemporary concerns. (The last flickering of this dying tradition can be seen in the recent work of the very oldest of distinguished scientists, such as Ernst Mayr and Ilya Prigogine.) What has happened in the interim to alter this situation?

As with all things involving the natural sciences, one can always find Dr. Pangloss ready to be heard. Following Elizabeth Eisenstein, we might say that scientists do not need to repeat, in any great detail, the histories of their fields because such accounts are now readily available on the shelves of any major library or bookstore, testifying to the success of the history of science as a publishing venture.[10] The attraction of a single synoptic text that brought the reader from a field's earliest history to its latest developments quickly faded with the economic feasibility of a personal library. Thus, once freed from recapitulating the past, scientific authors were left with more time to push back the frontiers of knowledge. Of course, this story will appear persuasive if one sees the relevance of history to contemporary science exclusively in pedagogical terms, namely, as hastening the reader's intellectual progress until she is on top of the latest developments. Conveniently omitted, however, is the idea that history might be a resource that bears *critically* on contemporary issues. After all, a division of labor is 'functional' only if prior agreement has been reached on the sort of product that the workers must produce.

At this point, we may introduce a more 'Faustian' explanation for why scientists have retreated from philosophically sophisticated and archivally responsible history. In fact, it is the creation myth that historians of science frequently tell about their own field, the most famous version of which may be found in Kuhn. The basic idea is that contemporary irrelevance is the price that historians have had to pay for retaining their autonomy.[11] Assumed here is that this bargain was the outcome of a hard-fought battle, whereby historians wrenched away the possibility of an objective historiography of science from interested and powerful scientific parties. The most aggressive version of this story portrays the history of science as saving science from itself, specifically by preserving its culturally ennobling features in a world where scientific work has become increasingly barbaric in both ends and means. This sensibility, one aligned with 'humanist' approaches to general cultural history, can already be detected in George Sarton, the founding editor of *Isis*, after he emigrated from Belgium to Harvard during World War I.[12]

The only problem with this scenario is that it presupposes that the natural sciences are fit objects of humanistic inquiry — albeit the 'final frontier' of the humanities. However, before the Cold War, there is little evidence that many humanists held such a view. Clues here may be found in the persistent separation of 'history of science' from 'history' departments in universities. Nineteenth century humanists discussed the superiority of the natural sciences as an outgrowth of the superiority of European culture. Thus, Galileo was presented as an expression of Renaissance genius, Newton of Enlightenment genius, and so forth. But as soon as it became clear that the natural sciences could flourish even in the hands of non-Europeans — especially the Japanese after their 1905 War with Russia — the cultural specificity of science, so crucial for humanistic inquiry, was shattered.[13] The point was driven even more forcefully by the fact that the Japanese were quite deliberate in seeking mastery only in some European forms of knowledge, but not others. All told, these developments called into question the need to be 'European' in order to be 'scientific'.

The other historically important obstacle to assimilating the study of the natural sciences to the humanities has been science's reliance on manually operated technology, especially laboratory apparatus, for its inquiries. The unity of research and teaching — the hallmark of Humboldt's influential model of the modern university — was to be achieved by a common interest that researchers and students would have in books. However, the natural sciences have depended on book-learning almost entirely for educational, but not research, purposes. (In fact, this distinction has contributed to the unravelling of the Humboldtian ideal in the twentieth century). Consequently, these fields held no special respect for the literary forms of knowledge found in the archives, except as vehicles for the 'virtual witnessing' of natural or experimental phenomena. Indeed, this antipathy to the ethos of 'the liberal arts' (which were literally free from manual involvement) partly accounts for why two hundred years had to pass before the Scientific Revolution was fully institutionalized as proper university subjects.

Because the natural sciences were ill-suited to humanistic inquiry, strategies had to be devised to render them proper objects. Two stand out for their influence on contemporary history and sociology of science. The first, associated with Ludwik Fleck, assimilates the understanding of the laboratory sciences to models derived from art history.[14] This was a natural move for Fleck himself to make, given the historical ties between his own science of medicine and the artistic professions. Thus, via Fleck's concept of 'thought-style', one can witness the metamorphosis of artistic schools into scientific paradigms.[15] The legacy of this approach is felt in contemporary studies that highlight the iconographic aspects of laboratory practice (i.e. its appropriation of multiple scientific traditions), often to the neglect of the ambient political economy that sustains the practice. The second strategy, associated with Alexandre Koyré,[16] simply subordinates laboratory experiments to the scientific discourses in which they figure as evidence. Koyré's is essentially an updated version of the seventeenth century philosophical view that experiments are little more than heuristics for enhancing one's dialectical and hermeneutical reasonings. And while contemporary sociologists of science would be the last to want to be associated with Koyré, nevertheless more than one practicing scientist has remarked that only a dyed-in-the-wool humanist would ever view a laboratory as an elaborate 'inscription device', as Latour and Woolgar do.[17]

Together these two strategies have given the history of science much of its 'autonomous' character, which in turn has set it apart from other branches of history. Indeed, science seems to have an internal sense of development *only* when historians of science write about it. As part of a general history of the West, it is more common for science to be portrayed as a manifestation of ambient cultural forces. To exaggerate the point slightly, a historian of science will understand Newton diachronically in terms of his scientific precursor Galileo, whereas a historian is more likely to understand him synchronically in terms of his non-scientific contemporary Locke. And while, of course, this distinction in approaches is slowly being eroded, it is nevertheless striking just how often the names of academic disciplines are still used to organize historical inquiry into science, thereby conveying an air of long-term continuity to the object under study. I am as guilty of this as anyone, but an especially vivid case was provided in Frederic L. Holmes' keynote address to the Göteborg Workshop, "Writing about scientists now and in the past".[18]

In both his address and subsequent remarks, whenever Holmes discussed the main obstacles facing historians of contemporary science, he invariably fell back on disciplinary — rather than class, national, or even generational — differences as being the most salient ones dividing historian and scientist. Such a move reflects a strong bias toward the culture of the university. Admittedly, this is where humanists are most likely to encounter scientists. Yet, it has been only over the last hundred years that universities have played a preponderant role in the advancement of natural science. Moreover, that may turn out to have been a temporary arrangement, if the current calls to privatize, or at least de-academicize, scientific research are heeded. Indeed, a common theme in the Göteborg Workshop concerned the problems that historians had in making sense of the norms surrounding the production and distribution of knowledge in non-university settings. In particular, many found it hard to pin down the exact role that science plays in government agencies and industrial firms, places where scientific research is heavily supported, yet where the scientist is not primarily focussed on contributing to a discipline's theory and practice.

Given the allure of the Faustian fantasy described earlier, it would be a rude awakening, indeed, if professional historians of science were to learn that many leaders of the scientific community have been quite happy to see historical matters hived off from their fields and rendered into fit objects of humanistic inquiry. At the very least, the autonomous view of science promoted in such histories is sure to contain whatever delegitimatory effects might result from surprises unearthed in the archives. Perhaps because we are still under the spell of the logical positivists and their Popperian cousins, this point is easily missed.

When one speaks of 'criticism internal to science', we are likely to think of 'crucial experiments' and other empirical findings that scientists have made in laboratories and other sanctioned research settings. While historical research may be relevant for 'external' critiques of science, it is not normally seen as one of the avenues open for practicing scientists to influence the day-to-day activities of their fields. As noted above, the political advantages of marginalizing the historical record in this fashion are clear. However, the methodological advantages are not so immediately apparent. One would have thought that the historical record — precisely because it is constructed by agents who have a limited understanding of their own motives, audiences, and impacts — would be a source of 'naturally occurring' phenomena against which to test scientific claims that purport to build on that record.

Suppose that a distinguished scientific progenitor overplayed the weakness of rival positions and used his influence to avoid answering damaging objections to his own position. Why shouldn't his present-day intellectual heirs be held accountable to this record, instead of presuming that there is a 'statute of limitations' on how long one is required to take criticism seriously? If you find this line of reasoning fanciful, just remember that it was precisely the one used by Ernst Mach in *The Science of Mechanics* to revive the early objections to Newton's postulation of absolute space and time,[19] which later inspired Einstein to construct The Special Theory of Relativity. For a sense of our distance from this sensibility, consider the difference in audience and impact of Paul Feyerabend's self-consciously Machian reassessment of Galileo's scientific contributions.[20] While it had some influence on philosophical attitudes toward the history of science, it failed to alter any major scientific research agendas.

Perhaps more than we care to admit, science has internalized the Social Darwinist ethic implied in what Kuhn has called the 'Planck Effect': If you outlive — or out-reproduce — your critics, you've *earned* your right to forget what may have been a sordid past.[21] Sometimes, the point is made more delicately by saying that 'philosophically inspired' criticism is beside the point once a research tradition has solved enough of the empirical problems it has set for itself.

At that point, criticism that originally appeared 'foundational' is now merely 'external' to the proper business of science.

THE EPISTEMIC ROLE OF HISTORY IN SCIENCE VERSUS THE HUMANITIES

So far, I seem to be suggesting that historians can function only as epistemic referees, ensuring that credit is given where due and criticism addressed where posed. But aside from maintaining methodological standards, historians are also in an excellent position to recover contingent features of scientific practice that are mistaken for necessary ones, a conflation (one might argue) that only serves to retard natural science's aspiration to being a 'universal' form of knowledge. Such conflations are routinely made whenever scientists identify the pursuit of a research programme with the use of specific methods, techniques, and instruments. Presumably, if 'science' is indeed separable from the scientists who practice it at any given time, then it would be a mistake to equate the advancement of science with the perpetuation of the skills and competences of any particular group of scientists. (But of course, it would be a mistake that would be in many scientists' interest to make!)

For example, when critics of high-energy physics suggest that it would be more cost-effective for physical theories to be tested on computer simulations than on ever larger particle accelerators,[22] physicists treat the suggestion as an attempt to let extraneous financial considerations prevent them from getting at the nature of reality. Yet, particle accelerators are themselves devices for simulating the first milliseconds of the universe after the 'Big Bang' is supposed to have occurred. Not surprisingly, these instruments were embroiled in controversy over their own kind of 'artificiality' when they were first proposed a half century ago. Needless to say, the identification of historical precedents alone will not be sufficient to change physicists' minds, but at least the argument can return to what is really at stake, namely, whether the current crop of physicists have the computational skills needed to make the proposed shift in testing methodology.

As one surveys the academic disciplines, it quickly becomes clear just how unique the natural sciences are in excluding history from strong epistemic roles. In contrast, throughout the humanities and social sciences, historically based arguments have played major roles in theoretical debate. Indeed, an alternative historical scenario underwritten by some neglected facts has been known to reverse burdens of proof and upset balances of power in these fields, often leaving the outsider with the impression that the humanities and social sciences are permanently unsettled, or undergoing 'permanent revolutions'.

Of course, one should not exaggerate the success of today's historians of the social sciences in influencing the course of their fields. There is no doubt that, broadly speaking, these fields have acquired many of the institutional trappings of the natural sciences (especially after Thomas S. Kuhn showed them how), and so historical courses of study are increasingly treated as marginal to the state of disciplinary play. Nevertheless, the *aspiration* of historians of the social sciences to change their own fields is still very much present. Otherwise, why are historians of natural science more likely to see the History of Science Society as their main professional body than historians of social science, who still claim primary allegiance to the discipline whose history they study? Also, constructivism acquires a much more apocalyptic tone in histories of social science than of natural science. In the former case, not only is there a sense that things could have been otherwise, but also that things can still be difference, once we come to accept a different account of the past as our own.[23]

Moreover, it would be a mistake to see this apocalyptic sense of constructivism — whereby an alternative sense of the past promises to deliver an alternative sense of the future — as just so much postmodern decadence. On the contrary, the sensibility was widespread when historical inquiry was considered the primary mode of scientific criticism. This was during the Enlightenment, when the paradigm was the 'critico-historical' method in theology, with its two-pronged strategy of demystifying superstition followed by rationally grounding the demystified domain: myth replaced by science.[24] The German critic Gotthold Lessing put the point most incisively toward the end of the eighteenth century, when he defended himself against religious censorship by arguing that the ultimate trial of faith is to continue believing, in spite of being shown that all the concrete accompaniments of faith are false.

Lessing was clearly trying to portray the Enlightenment as a replay of the Reformation by alluding to the flattering self-image of the Protestants as having seen through the ecclesiastical materialism of the Catholics. More importantly, regardless of what one makes of Lessing's sincerity, he certainly turned out to be correct in claiming that a critical attitude toward religious practice, however much it may threaten religious authority, did not undermine religious faith as such. Over two centuries after the emergence of critico-historical theology, Christianity is still very much with us as a 'world religion', albeit in somewhat more secularized forms. It is important to keep this point in mind, given the sorts of arguments that natural scientists — and even such historians of science as Kuhn and Stephen Brush — have made for excluding historically based criticism from the professional training of scientists.[25] Should we suppose that faith in science is so much more vulnerable to historical criticism than faith in God? Would science wither in the face of its own past?

From the testimony of Max Planck and Ernst Mach during their famous exchanges over the ends of science in the first decade of this century, one would have to conclude that the answer to both questions is 'yes'. These exchanges took place at a turning point in the institutionalization of science. After a generation of being securely housed in the universities, the natural sciences had developed research trajectories that increasingly diverged from the instrumental ends that had led the state and industry to support scientific research in the first place. Mach saw this narrowing of science's horizons as indicative of dogmatic closure that undermined both the spirit and the utility of scientific inquiry. His remedy was to demystify the tale of paradigm-driven progress by recovering traditions of inquiry that have not participated in the construction of the dominant *Weltbild*. His efforts of recovery ranged from demonstrating how *Naturphilosophie* could explain the folk science of rural Austrian artisans to supporting Buddhist resistance to the imposition of a British scientific curriculum in Ceylon.[26]

Both Mach and Planck were self-conscious of the precedents from theology.[27] Mach had subtitled his major philosophical critique of contemporary physics, *The Science of Mechanics*, a 'critical history'. Moreover, he gladly accepted the mantle of 'heretic' in the face of Planck's own avowed role of 'high priest'. According to Mach's sense of history, the past is always being carried into the present, and hence always a resource for altering the future. For his part, Planck openly admitted — in a way that would embarrass his fellow scientific realists today — that convergence on an ultimate reality requires a common mindset at the outset of inquiry. The uncertainties of the archives are thus anathema to the realist formula. Dogma, by contrast, has a unique ability to focus the mind.[28] The past is best left there, reasoned Planck, as the surest bet for the future is to continue with what seems to work in the present.

Despite its long-term role of removing history from the arena of science criticism, the Mach-Planck controversy was, in its own day, interpreted as isomorphic to a common debate in critical-historical theology. Does one grasp Christ's message by looking at what is distinctive about Christianity or what Christianity shares with the other world religions? Similarly, does one grasp the ends of scientific inquiry by looking at what is distinctive about contemporary natural science or what it shares with other knowledge producing practices throughout the world? In opting for the former, Planck stressed science's unprecedented degree of discipline and focus, whereas Mach, in choosing the latter, emphasized science's role in enabling people to economize on effort in whatever they wished to pursue. This difference of opinion turned out to be quite important in debates over the role of science in general education, with Planck's view ultimately prevailing.[29]

Once the natural sciences were ensconced in the academy, they soon came to set the standard by which the other disciplines were judged. What made them impressive was the amount of control they exercised in the environments where they operated. I do not mean here that natural scientists 'control' natural phenomena better than social scientists 'control' social phenomena. (In fact, no one has ever shown this to be the case.) Rather, I mean that natural scientists control *each other* better.[30] Specifically, their acts of collective remembering are tightly constrained. Kuhn aptly calls this phenomenon "Orwellian".[31] For example, the natural sciences have higher journal acceptance rates because their practitioners know that the articles expected of them should cite the latest, not the earliest, statement of a theory, method, or finding. In that way, the illusion of a foreshortened past accelerating toward the present is maintained.[32]

The only recognizable role for a historian in this exercise is in the adjudication of originality: Everyone wants to take credit for having made the same discoveries. Thus, historical interest becomes narrowly focussed on who got 'there' first rather than on how many other ways one could have gone.

While casting out the historians from the temple of science may make it easier for researchers to consolidate around a common image of linear progress, that was not the only consideration that led Planck to denounce Mach as a heretic. A more pressing concern was that science remain an autonomous voice that dictated the terms on which it dealt with the state and industry.[33] If history shows that the ends of science have been dictated by concrete human needs, as Mach maintained, then that would seem to imply that science is a pure instrument at the behest of whomever speaks for humanity at a given time. Here it is worth keeping in mind that, in the first decade of the twentieth century, the main ideologues of natural science, the Positivists and the Pragmatists, tended to follow Mach in blurring such distinctions as science vs. technology, theory vs. application, truth vs. utility. Planck's call for autonomy was very much in the minority at the time, yet his worst fears were vindicated by Germany's humiliating defeat in the World War I, an effort that was fueled by the enthusiastic participation of Mach's strongest supporters, the chemists. Not surprisingly, the 1920s witnessed the emergence of several philosophical attempts — the Vienna Circle's being the most notable — to portray science as an autonomous truth-oriented, theory-driven enterprise, quite separate from technology and any of its effects.

The pattern of Planck's concerns should be familiar to anyone who has been following the recent Anglo-American debates over 'science's social standing' between natural scientists and the sociologists who study them.[34] Public remarks about the context-relative, interest-driven character of scientific research may appear, at first glance, to be politically innocuous. But to a society that has invested much of its material and symbolic resources on such research, and has already begun to question the return on that investment, these remarks are received as a call for renegotiating science's social contract. Leading scientists fear that their research budgets will be either cut entirely or (perhaps worse from their standpoint) subsumed under current political, social, and economic agendas. Unfortunately, the best that these scientists have been able to muster in response is a watered-down repetition of Planck's Orwellian strategy of promoting the myth of progress while denying the relevance of history and sociology to the science curriculum. Still more unfortunately, historians and sociologists have so far done little to puncture the myth in the public sphere. The main reason, I submit, is that we have yet to challenge the myth on its own terms with an alternative 'grand narrative' of the history of science — especially one that scientists can see as their own. (Of course, there need not be just one such alternative, but one would do for a start.) At the moment, the case study method currently preferred by professional historians and sociologists carries the rhetorical force of isolated counterexamples, if not mere nitpicking, once translated into the public forum.

THE NEED FOR GRAND NARRATIVES OF THE HISTORY OF SCIENCE

As it turns out, there is a model for an alternative grand narrative of the history of science, though it is one that would need to be considerably refurbished before having contemporary purchase. It was most explicitly articulated in John Desmond Bernal's four-volume *Science in History*.[35] The title of the project is itself revealing, as it already shifts the emphasis from the idea of science having its own history to that of science being embedded in general history. Also interesting is the fact that Bernal was not a humanist, but an X-ray crystallographer who believed that contemporary images of scientific progress were grounded in emerging relationships between scientists and the state and industry that were, over the long haul, detrimental to the integrity of science.

According to Bernal, as the ranks of scientific labor swelled in the twentieth century, an increasing number were being siphoned off into defense-related industries, the net effect of which was to contain the destabilizing tendencies that scientific innovation has historically had on existing power relations. This trend could be seen in the increasing fragmentation of the scientific community, euphemistically called 'specialization', which is driven less by a functional differentiation of subject matter and more by the divisions of labor imposed by the logics of capital accumulation and nation-building. As is well known, Bernal wanted to raise the class consciousness of scientists by persuading them that, while separately they may be contributing to destructive societal tendencies, united they can reverse those tendencies by demanding a distinct voice in the deployment of their expertise.[36]

Given what was said earlier about the Mach-Planck controversy, the interesting feature of Bernal's strategy is its reliance on a critical history of contemporary science as propaedeutic, not antagonistic, to an assertion of scientific autonomy. However, the historian need not be as strongly committed to a Marxist agenda as Bernal to convey a similar sensibility. Many people who eventually become professional historians of science were themselves originally planning to be scientists but then became disillusioned with the degree of specialization and competitiveness required by their chosen field. Under those circumstances, writing about the past becomes a way of keeping alive earlier models of inquiry which no longer obtain.

Indeed, this may even be said of The *Structure of Scientific Revolutions* and its author, a Ph.D. in physics who failed to make the transition from 'little' to 'big' science at the end of the Second World War.[37] For a book normally thought to provide a model of scientific change of contemporary import, it is striking that *Structure* contains no examples drawn after the revolution in quantum mechanics in the 1920s. (In the case of chemistry, the latest examples are from the mid-nineteenth century.) One reason may be the one cited at the start of this paper, namely, Kuhn's historiographical strictures against studying contemporary science. But a deeper reason may be that the logic of puzzle-solving, the focus of Kuhn's model of normal science, fails to explain the research trajectory of science in the contemporary period. In other words, contrary to the latter-day appropriations of his work, Kuhn's model is specific, roughly to the period 1620 to 1920. It is only a small step from defining the historical parameters of the model to using it as a normative basis for criticizing contemporary science for failing to live up to that model. As a tighter labor market drives aspiring scientists into such non-scientific careers as the history of science, I envisage that this step will be increasingly made.

NOTES

1 Kuhn 1977, 3–20, esp. 16.
2 Fuller 1992.
3 Kuhn 1978.
4 Kuhn *et al.* 1967.
5 Söderqvist, (forthcoming).
6 Collins, "The unwritten lore", unpublished paper for the Göteborg Workshop.
7 Hughes (this volume).
8 Laudan 1993.
9 Kuhn 1970.
10 Eisenstein 1979.
11 Fuller 1994c.
12 Sarton 1948.
13 Pyenson 1993.
14 Fleck 1979.
15 Cf. Kuhn 1977, 340–52.
16 Koyré 1957.
17 Latour and Woolgar 1979.
18 Holmes (this volume).
19 Mach 1960.
20 Feyerabend 1979.
21 Kuhn 1970, 151.
22 Lindley 1993.
23 Mirowski 1989, Danziger 1990; Wagner *et al.* 1991.
24 Brooke 1991, Ch. 7.
25 Kuhn and Brush 1974.
26 Blackmore 1972, 206–7; 1992, 185.
27 Fuller 1994d.
28 Cf. Kuhn 1977, 225–39.
29 Fuller 1994a.
30 Fuchs 1992.
31 Kuhn 1970, 167.

32 Fuller 1994b.
33 Heilbron 1986.
34 Johnson 1995.
35 Bernal 1971.
36 Jones 1988.
37 Kuhn 1970.

BIBLIOGRAPHY

Bernal, J.D., *Science in History*, 2 vols. (Cambridge: MIT Press, 1971).

Blackmore, J., *Ernst Mach* (Berkeley: University of California Press, 1972).

Blackmore, J. (ed.), *Ernst Mach: A Deeper Look* (Dordrecht: Kluwer, 1992).

Brooke, J.H., *Science and Religion: Some Historical Perspectives* (Cambridge: Cambridge University Press, 1991).

Brush, S., "Should the history of science be rated X?", *Science*, vol. 183 (1974), 1164–1183.

Danziger, K., *Constructing the Subject* (Cambridge: Cambridge University Press, 1990).

Eisenstein, E., *The Printing Press as an Agent of Change* (Cambridge: Cambridge University Press, 1979).

Feyerabend, P.K., *Science in a Free Society* (London: New Left Books, 1979).

Fleck, L., *Genesis and Development of a Scientific Fact* (Chicago: University of Chicago Press, 1979).

Fuchs, S., *The Professional Quest for Truth* (Albany: SUNY Press, 1992).

Fuller, S.W., "Being there with Thomas Kuhn: a parable for postmodern times", *History and Theory*, vol. 31 (1992), 241–275.

Fuller, S.W., "Towards a philosophy of science accounting: a critical rendering of instrumental rationality", *Science in Context*, vol. 7 (1994a), 591–621.

Fuller, S.W., "Social psychology of science: another strong programme", pp. 162–178 in W. Shadish and S.W. Fuller (eds.), *Social Psychology of Science* (New York: Guilford Press) (1994b).

Fuller, S.W., "Teaching Thomas Kuhn to teach the cold war vision of science", *Contention*, vol. 4 (1994c), 81–106.

Fuller, S.W., "Retrieving the point of the realism-instrumetalism debate: Mach vs. Planck on science education policy, pp. 200–207 in D. Hull, M. Forbes, and R. Burian (eds.), *PSA 1994*, vol. 1 (East Lancing: Philosophy of Science Association, 1994d).

Heilbron, J.L., *Dilemmas of an Upright Man: Max Planck as a Spokesman for German Science* (Berkeley: University of California Press, 1986).

Johnson, J., "Science friction", *New Statesman* (January 13, 1995), 29–30.

Jones. G., *Science, Politics and the Cold War* (London: Routledge, 1988).

Koyré, A., *From the Closed World to the Infinite Universe* (Baltimore: Johns Hopkins University Press, 1957).

Kuhn, T.S., *The Structure of Scientific Revolutions* (2nd ed., Chicago: University of Chicago Press, 1970).

Kuhn, T.S., *The Essential Tension* (Chicago: University of Chicago Press, 1977).

Kuhn, T.S., *Black Body Theory and Quantum Discontinuity, 1894–1912* (Oxford: Clarendon Press, 1978).

Kuhn, T.S., J.L. Heilbron, P. Forman, and L. Allen, *Sources for the History of Quantum Physics* (Philadelphia: American Philosophical Society, 1967).

Latour, B. and S. Woolgar, *Laboratory Life: The Social Construction of Scientific Facts* (London: Sage, 1979).

Laudan, R., "Histories of science and their uses: a review to 1913", *History of Science*, vol. 31 (1993), 1–34.

Lindley, D., *The End of Physics* (New York: Basic Books, 1993).

Mach, E., *The Science of Mechanics* (6th ed., La Salle: Open Court Press, 1960).

Mirowski, P., *More Heat Than Light* (Cambridge: Cambridge University Press, 1989).

Pyenson, L., "Prerogatives of European intellect: history of science and the promotion of Western civilization", *History of Science*, vol. 31 (1993), 289–315.

Sarton, G., *The Life of Science: Essays in the History of Civilization* (New York: Henry Schuman, 1948).

Söderqvist, Thomas, "After the 200th Hour: The A/effects of long-term interviewing for science biography", in Horace, F. Judson and Thomas Söderqvist (eds.), *Interviews in Writing the History of Recent Science* (Cambridge, MA: Harvard University Press, forthcoming).

Wagner, P., B. Wittrock, and R. Whitley, *Discourses on Society: The Shaping of the Social Science Disciplines* (Dordrecht: Kluwer, 1991).

Index